Ecologies of Inequity

SERIES EDITORS

David L. Brunsma
David G. Embrick

SERIES ADVISORY BOARD

Margaret Abraham
Elijah Anderson
Eduardo Bonilla-Silva
Philomena Essed
James Fenelon
Tanya Golash-Boza
David Theo Goldberg
Patricia Hill Collins
Evelyn Nakano Glenn
José Itzigsohn
Amanda Lewis
Michael Omi
Victor Rios
Mary Romero

Ecologies of Inequity

How Disaster Response Reconstitutes Race and Class Inequality

Sancha Doxilly Medwinter

The University of Georgia Press
ATHENS

© 2023 by the University of Georgia Press
Athens, Georgia 30602
www.ugapress.org
All rights reserved
Designed by Kaelin Chappell Broaddus
Set in 10.5/13.5 Garamond Premier Pro Regular
by Kaelin Chappell Broaddus

Most University of Georgia Press titles are
available from popular e-book vendors.

Printed digitally

Library of Congress Cataloging-in-Publication Data

Names: Medwinter, Sancha Doxilly, author.
Title: Ecologies of inequity : how disaster response reconstitutes race and
class inequality / Sancha Doxilly Medwinter.
Description: Athens : The University of Georgia Press, [2023] | Series: Sociology of race and ethnicity | Includes bibliographical references and index.
Identifiers: LCCN 2022054493 (print) | LCCN 2022054494 (ebook) | ISBN 9780820363813 (hardback) | ISBN 9780820363806 (paperback) | ISBN 9780820363820 (epub) | ISBN 9780820363837 (pdf)
Subjects: LCSH: Disaster relief—Moral and ethical aspects—United States. | Disaster victims—United States. | Discrimination—United States. | Classism—United States. | Racism—United States.
Classification: LCC HV555.U6 M46 2023 (print) | LCC HV555.U6 (ebook) | DDC 363.34/80973—dc23/eng/20230513
LC record available at https://lccn.loc.gov/2022054493
LC ebook record available at https://lccn.loc.gov/2022054494

To the Superstorm Sandy disaster survivors, thank you for sharing your stories with me. By the time you find yourselves in this book, I trust that you have found a way to gather up the pieces and put them together as best as you could. To my family, Floyd, Pristeene, Melinda, Pauliciana, Priscilla, Ma Ju, Defwen, Titus, Abi, Ursula, Martin, Roger, and Juliette, you taught me how to love and see the humanity in people, always wanting the absolute best for them. You will continue to inspire all my future books as I align my career with my calling, my dedication to understanding and undoing inequality, inequity, and injustice.

CONTENTS

ACKNOWLEDGMENTS ix

PROLOGUE xv

INTRODUCTION 1

CHAPTER 1. Ecologies of Inequity 15

CHAPTER 2. Race-Class Logics of Urban Spaces 23

CHAPTER 3. Black Immigrants and Disaster Inequality 31

CHAPTER 4. Labyrinth Bureaucracy 49

CHAPTER 5. Social Capital in Crisis 67

CHAPTER 6. Logic of Response versus Services 79

CHAPTER 7. Social Capital Privilege 100

CHAPTER 8. Organizational Networks of High and Low Capital 114

CONCLUSION 131

EPILOGUE 143

APPENDIX A: INTERVIEW GUIDE 147

APPENDIX B: REFLECTIONS 151

APPENDIX C 155

REFERENCES 157

INDEX 167

ACKNOWLEDGMENTS

> Writing a book is seldom about placing words on a page. Authoring this book has been a journey, a journey with my village.
>
> —Sancha Doxilly Medwinter, the author

To my village, who began this walk with me, some who joined earlier, some later, some for short periods, some longer, some who planned to tarry but have transitioned to the other life, before ever having the chance to run patient fingers through the leaves, crisp and new, feeling their sharp edges before slowly tracing letters forming their names with fingertips, calloused from tilling the rich soil that led to plentiful harvests of produce that made a belly full, before having the chance to clasp with rope-burned palms from pulling back the cows too eager to run off down the hill sure to end delicious milky breakfasts, and possibly a Christmas, well-planned, were it not for those strong hands that were gentle enough to cause a child to melt, like cocoa sticks brought to boil in a charred-bottom cast iron pan, into the warm embrace of wrinkly arms, soft pillows made of love that ripples through generations, my heart thanks you.

To my husband Floyd Medwinter, for whom I have a deep love, I am so grateful for you. You have always lovingly and steadily supported my personal dreams and professional goals even when the evidence of their possibility was not found. I am grateful for your quick thinking to request from your employer the truck that would hold and transport the disaster supplies to the Superstorm Sandy survivors in such a short window. Thank you also for being a constant reminder and motivator to keep making progress on my writing.

I am so grateful for my girls, Melinda and Pristeene, who at eleven and twelve excitedly made a list of supplies and posted the flyers on the walls, stairwells, mailboxes, and doors of our apartment complex. Your work helped bring these families needed supplies that they had not yet received twelve days into the disaster. Keep considering how you may help the least fortunate and anyone who is experiencing tough times. I love and thank you young women, now twenty and twenty-one, for always looking up to me with an admiration that only daughters possess for their mothers.

My loving daughter Pristeene, you took this journey with me, a young single mom who made ramen and canned mixed vegetables from a rented bedroom with no windows, which meant we had to keep the door ajar to overcome the brutal New York City summers. Thank you for understanding why I always seemed busy juggling mothering, school, and work, for the better future that seemed forever on the horizon, all the while that I feared that I was failing you as a mother. Even at six years old, you believed my stories that we would have a Cinderella transformation and have a home where there would be enough room for you to spread out your toys and play. Thanks to my friend Princess, who first dreamed this dream for us, when I would bring you over to her apartment to get a sense of what a full and spacious home that was ours would feel like. Melinda, you're my special gift from a mom who entrusted me with her heartbeat even as hers was fading. You are my loving, bubbly daughter who could always sense when I needed a boost when you asked me how the book was going, and I responded with, "It's going." Your "You got this, Mom!" infused me with the energy I needed to continue.

My heart overflows with rivers of gratitude to my grandmother Mary Darius, who raised me from the age of one and cultivated in me a devotion to my studies. Always knowing that I was going further than most would have imagined. There would be no play until I completed homework and studying. Thanks to Priscilla Doxilly, my mother, and my queen, who birthed me into this world and whose unfailing love and support has carried me through college, then graduate school, and into my first job as a professor. You always spoke life into me. Thanks for telling me in moments of uncertainty that plagued my graduate school experience, "Sancha, if everyone counts you out; *You*, don't count yourself out!" No one knows this book as intimately as you do. Thanks for listening to my working theories, asking me important questions to help me dig deeper, accompanying me on one of my return visits to the field sites. You have always surrounded me with fasts and prayers that helped me through difficult periods in my career. I am grateful for my dad, Titus Emmanuel, a people's person. You have generously shared your wisdom on how to interact with people whom I encounter in my life's journey.

Mum and Pap, where do I begin? I will forever love and hold you dear to my heart. Mum (Ma Ju), and Pap (Defwen), my great-grandparents only whose spirits are still with us, but who gave me all that they had, a U.S. $100 bill, folded into a million halves wrapped in just as many plastic bags, much like Russian dolls. Yet you believed in me so much that you gave it to me in the wee hours of the morning in the few minutes we had left before catching my flight into the unknown. I didn't yet fully understand how precious that moment was when I hugged Pap, as it would be my last, a moment I would circle back to again and again through the years. I am so grateful I saw Mum again. She didn't recognize me at first, as I was no longer a teenager filled with dreams and the bravery to strike out on her own. I remember how your eyes welled up with tears as you called me my endearing name, "Suuuunnnnnnyyyy" with an endearment and longing that only my *sweet Ma Ju* could ever express. I am so grateful that I had the chance of that reunion.

My cheerleaders, my cousins, Juliette, Roger, and Martin. Roger and Martin, who would always ask me whether I had already finished the book, yes, I have! My cousin Juliette who has, for years, been plotting with my grandmother to parade this book in the communities in St. Lucia and Martinique as evidence that a *Doxilly* has made it to the highest rank in education and is now a published author, this book is yours to share. Abi and Sister Bastien, thank you for always making the time to celebrate with me. You are family. Tarrick, Nicky, and Nezry, thank you for accompanying me, and sometimes while I was interviewing someone, your friendliness kept others engaged, which sometimes gave me an opportunity to not miss them. Your knowledge of the lay of the land was priceless, and just having a travel buddy in the beginning made it less intimidating to go to the field alone.

My tribe, you who have kept me grounded, as I understand that my success is not mine only, but the success of many generations. You who have known scarcity, lacked access to education, but carried on a hope that in your lifetime you would see your dreams fulfilled in your great-grandchildren, grandchildren, and children.

This village who raised me and taught me how to love and see the humanity in people and always want the absolute best for them. I learned empathy from you. I learned how to love not just my family and friends, but everyone I encounter with a generous heart. These values that you have instilled in me have played no small part in fueling my passion and commitment to pursuing and completing this project, despite the emotional toll of encountering human suffering. You will continue to inspire all my future community-engaged projects, as I align my career with my calling, a dedication to understanding, so that I can help undo inequality, inequity, and injustice.

Robert, how can I ever forget that you are the one who gave me an opportunity to pursue my undergraduate degree and drilled into me and many other Black students, "You gotta get your PhD, you gotta get your PhD." You knew what I needed to fulfill my life's purpose. Carolina Bank-Munoz, my undergraduate mentor and member on my dissertation committee, you have been such a steady fixture at every stage in my career. Thank you for advising me and guiding me through my graduate school applications. I still remember the disbelief in your voice saying, "You're not going to Berkeley? No one turns down Berkeley!" I think I can safely say that you think I still turned out okay.

Thank you to my graduate school advisers and mentors who were instrumental in helping me accomplish the project, giving me feedback along the way. I am especially grateful to Nan Lin, who agreed to chair my dissertation after I had knocked on a few doors with no luck. I had never taken a course with you, yet when I called, you scheduled me for lunch after your return from San Francisco in three weeks. Assigning me five books on social capital in the interim and seeing how well versed I was by our first meeting showed you my seriousness about working with you. I am also grateful for the funding I received to further this work, a fellowship from Duke's Sanford School of Public Policy and a grant from the National Science Foundation.

I do believe there's been an angel at every station I disembark. Moon-Kie Jung, thanks for being a constant support and advocate for me during my time at UMass, especially those times when I most needed the support. You and Caroline were often a source of encouragement. Joya Misra, thank you so much for being such a resourceful and dependable mentor who has been such a great support in guiding my career trajectory. Your thorough read and generous editing of several iterations of this book and your helpful feedback are indispensable. David Brunsma, thanks for seeing the potential in the first draft of this manuscript and suggesting that I submit my proposal to University of Georgia Press.

Thanks also to David Embrick for your role as series coeditor in moving the work forward. Thanks to Mick Gusinde-Duffy for being such an encouraging and easy to work with executive editor. You clearly believed in this work. Thanks to referees 1 and 2; this book could not have been what it is today without you. Thanks to Jim Elliot for seeing the value in this work and inviting me to share with your graduate students. Their engagement with this work has helped me think deeper about my analysis.

A special thanks to Whitley Plummer, my cartographer, who created such beautiful maps, graphs, and tables for this book. You possess such professionalism, and detail-oriented, analytical smarts, which allows you to deliver. Tannuja Rozario, always a steady support, thank you for your consistent willingness to read, edit, and assist me with reorganizing some chapters. Thank you also for your efficient library research to add suggested sources by reviewers for the chapters. We really work well as a team, especially during that late night mad dash before sending out the last draft. Whew! Joanna Riccitelli, you were such an efficient and amazing editor who helped ensure that the manuscript was consistently and correctly formatted. Altogether, this process would have been so daunting without the assistance of these graduate students, whom I appreciate so much.

My sincere gratitude to the numerous disaster response site managers, staff, and volunteers who gave me their time to help me gain clarity on the limitations of disaster response. Thanks to the church leaders who became disaster responders overnight. I appreciate you bestowing upon me such deep insight into the process. I especially thank Pastors Errance and Ward and Bishop Fabian (pseudonyms, as are the names of all the clergy, staff, volunteers, and survivors used in this book) for facilitating my entry into the field and having done this work on a shoestring budget, with only rented trucks and no warehouse to store supplies.

To all Superstorm Sandy responders, your dedication to your work was visible and commendable. So many of the pitfalls are structural and cultural, both invisible. I only hope that the findings in this book will help improve disaster response and allow you to see the process from multiple perspectives and levels that were, at that time, inaccessible to you while you were deeply engaged in the work at your respective stations. Many of you worked unforgivable hours in an unfamiliar place, and away from family, so thank you.

I may have saved the best for last. I have nothing but immense gratitude, respect, empathy, and hope for all the Superstorm Sandy survivors in Canarsie, Brooklyn, and The Rockaways. Thank you for entrusting me with your vulnerability. Know that in the years since we last spoke there were long periods when, for one reason or another, I felt like I could not write. But your words, like your lives, were like drums that pounded on my heart and pulsated through my temples, not letting me rest until I fulfilled my promise to you. I promised to tell all who would read this book what happened here. This was especially true for Canarsie disaster survivors since many had erred in thinking that Superstorm Sandy had left you unscathed.

Ricky, thank you for your persistence, and your insistence on your humanity even though the system has let you fall through the cracks far too many times. I am forever changed by your generosity in allowing me to learn about your struggles. Your lived experiences allow us all to gain a deeper understanding of the complexity of navigating poverty, social services, and disaster response in New York City. I deeply regret that I lost touch with you shortly after your relocation to Brooklyn from The Rockaways like a pawn in "the system" (as you called it!) that has repeatedly tried to strip you of your agency. I had truly hoped for a miracle, that you would really find that apartment, permanent housing to welcome your baby, just as you wished. I know that you have hopes and that you have dreams too, and that you have an unlimited supply of optimism and fight; even as I am just as certain that we will all continue to fail you, and so many others like you, until our conscience won't allow us.

I am grateful that there are pastors of small churches, and founders, staff, and volunteers at community-based organizations, and local community members who within a day had mobilized. You are the ones who remain when the disaster response machinery retreats. Freddie, the founder of Always With You (the names of all organizations mentioned in the book are pseudonyms), your work in Eastville and your reflections on Eastvillers (both are pseudonyms) demonstrate your deep understanding of and the love you hold for your community. As I reflect on the hardships, frustration, confusion, disbelief, despondence, fatigue, anguish, and helplessness, I am also reminded of the rivaling, unrelenting will of local community.

I hope that this book will bring clarity, empathy, and resolve to those who are in positions of power and influence about the value of local community and homegrown organizations. These organizations serve in urban areas that the State socially, politically, and economically deprives. It is my hope that a deeper understanding of how we manufacture inequality will lead to needed allocation of resources, information, and respect to support your work, not only in the time of disaster but through the crises our communities routinely endure. The journey continues as we, with our village, continue to walk.

This book is derived in part from an article published in *Environmental Sociology*, 2021 © Taylor & Francis, available online: http://www.tandfonline.com/ DOI : 10.1080/23251042.2020.1809054.

Funding Sources

The project that culminated into this book to receipt of a grant from the National Science Foundation (SBE DDRIG # SES-1434602), and a fellowship from The Center for the Study of Philanthropy and Voluntarism at Duke University's Sanford School of Public Policy.

PROLOGUE

The Racial Capitalist State

In order to fully grasp how race and class fundamentally shape governmental and nongovernmental disaster response, we need to start with a baseline understanding that the United States is a racial capitalist state. My framing of the United States as a racial capitalist state rests on the basis that historically to the present, the United States was built on and continues to be sustained on racial, ethnic, and class exclusion, exploitation, expropriation, and violence.

The United States racial capitalist state amasses wealth through racial capitalism. Cedric Robinson, in his book *Black Marxism*, coined the concept of racial capitalism to capture the historical relationship between global capitalism and racism. European capitalists came from different ethnic groups than did the peasants whose labor they exploited. Then European capitalists' expansionist goals led to a fulfilled desire to control entire regions and exploit populations, which were phenotypically non-White, non-English-speaking, and with unfamiliar cultures (Robinson 2000). Today, while we erroneously refer to those who fall outside the category of White as "racial minorities," I adopt the more fitting term "racially minoriti[z]ed," which was coined by Yasmin Gunaratnam (2003) in order not to lose sight of this fact of their subjugation.

As these systems of domination and exploitation through capitalism expanded, so did the need for their justification. Therefore, racial ideology developed to "exaggerate regional, subcultural, and dialectical differences into 'racial' ones" (Robinson 2000, 25). Robinson (1983, 2000) argues that racism and capitalism developed concurrently and codependently. In this historical context, European capitalists developed and perfected racial ideology as their tool of justification for instituting systems of slavery and servitude globally beyond Europe's borders (Robinson 2000).

However, we need to go beyond thinking of racial capitalism as merely a system of racial exploitation. It is also a system of racial violence, domination, expropriation, and exclusion. Nancy Fraser (2016) critiques exploitation-centered conceptualizations of racial capitalism for their lack of accounting for the difference between the "exploitable citizen-workers and [unfree and] dependent, expropriable subjects" (163). Historically to the present, the expropriated subjects are the "chattel slaves, indentured servants, colonized subjects, 'native' members of 'domestic dependent nations,' debt peons, felons, and 'covered' beings, such as wives and children" (165). Today the expropriated subjects are the descendants of these classes, many of whom live in cities within the United States (Fraser 2016).

The racial capitalist state is a modern empire state, built and sustained on racial violence on racialized others (Jung and Kwon 2020, 1014). Moon-Kie Jung and Yaejoon Kwon conceptualize the racial empire state as a colonial, imperial state, consisting of hierarchies of colonized and noncolonized peoples and "incorporated" and "unincorporated colonial spaces" (1014). Genocide of Native Americans and what Orlando Patterson describes as the "permanent, violent domination" or "social death" of African Americans through slavery marked the birth of this empire (Jung and Kwon 2020, 1015; Patterson 1982, 13).

The racial capitalist state excludes those who occupy the lowest rung of racial and class hierarchies, through the denial of social and legal citizenship rights. The United States, through a history of engagements of usurping sovereign nations and territories, has political subjects varying in rights and citizenship and characterized by amorphous colonial spaces (Jung and Kwon 2020). Jung and Kwon (2020) foreground the racial empire state's use of control, coercion, and violence by police forces against Black citizens and noncitizens as a key strategy to maintain order. Similarly, even the awarding of citizenship to Native Americans, Native Hawaiians, Puerto Ricans, as well as the status of resident alien to various immigrants, does not point to assimilation but to a means of racializing these groups (Jung and Kwon 2020). The key to the empire state is that it carries out similar policies of violence such as the "war on terror" on racialized subjects both on the mainland and beyond (Jung and Kwon 2020).

The racial capitalist state deprives those it deems excludable, exploitable, and expropriable from any real chance to survive and thrive. The racial capitalist state manufactures social, political, and economic deprivation through what Johan Galtung (1969) called "structural violence." The social structure manufactures lack of freedom, chronic poverty, and the psychological and physical suffering that stems from these suboptimal conditions for human life (Galtung 1969). The travesty is that this structural deprivation is preventable due to the availability of resources, but a lack of will among those who benefit from the status quo keeps this unjust machine running (Galtung 1969).

In favor of capitalist interests, the State enacts policies that pursue a neoliberal ideology that routinely erodes the social safety nets of the economically deprived. Neoliberal ideology holds that the market is the only regulator of social goods (INCITE 2020). The proponents of neoliberalism eschew governmental intervention in meeting the needs of the economically deprived. However, in practice, neoliberal policy requires "an active state to direct the dismantling of social welfare programs, the deregulation of labor and trade, and the protection of the wealth and assets of transnational corporations and a global elite class" (xiv). Milton Friedman and the Chicago school were the proponents of neoliberalism as a U.S. policy (INCITE 2020). In the 1980s, Reagan adopted neoliberal policies that dismantled labor organizing and allowed for unbridled racial capitalism (INCITE 2020). Reagan's neoliberal policies eroded social programs that were meant to combat the urban poverty that resulted from massive unemployment in deindustrializing cities.

The racial capitalist state pushes the precariously employed and unemployed Black, Latino, and noncitizen urban economically deprived into segregated urban residential clusters that are chronically disinvested by the State (Taylor 2014). On the other hand, the State may subsidize the cost of the means of production for the capitalist class (Fraser 2016). The State also ushers in "legal frameworks that legitimate the confiscation of the land and resources of 'dependent paupers,' convicted felons, undocumented workers, and colonial immigrants of color" (172). Most economically deprived residents, in the absence of housing assistance from the State, live precariously, which either ends in eviction or homelessness (Desmond 2016).

Economically deprived neighborhoods become a money pit for landlords who enjoy low taxes and low interest rates yet charge high rents to already economically deprived urban residents (Desmond and Wilmers 2019). According to Matthew Desmond and Nathan Wilmers (2019), "nationwide, the median rental unit located in an economically deprived neighborhood yields $98 in profits, compared to only $3 in middle-class neighborhoods and $49 in affluent neighborhoods" (1108). The annual rent in economically deprived neighborhoods far exceeds the property value (Desmond and Wilmers 2019). This rent to value ratio is much higher in economically deprived neighborhoods than it is in economically privileged neighborhoods. The economically deprived also pay a higher rent burden of 50–70 percent of their income (Eggers and Moumen 2010, cited in Desmond and Wilmers 2019, 1091 [Eggers and Moumen 2010]).

In northern cities, developers and landlords amassed great wealth by allowing "slum" conditions to grow in Black and economically deprived neighborhoods. One of their strategies is dividing dilapidated homes into smaller units without incurring costs of repair (Desmond and Wilmers 2019). Landlords exploit Black renters by charging high rents with no obligation to rehabilitate the old housing stock (Desmond and Wilmers 2019).

Global financial institutions, many of which are based in global cities such as New York City, also promote dispossession of economically deprived and working classes by debt, foreclosures, predatory loans, and other instruments of confiscation (Fraser 2016). In my fieldwork, I encountered the use of similar expropriation and exploitation tactics on Black, Latino, Native American, White ethnic, and undocumented impoverished residents on The Rockaway peninsula (henceforth "The Rockaways").

Nowhere else have I seen such a sharp contrast between the White protected class and the exploited, expropriated classes of the racial capitalist state than in the segregated landscape of The Rockaways. I spoke with Westvillers (pseudonym), who are White, economically privileged to affluent, and who reminisced about their interrupted leisure, which includes jogging along the beaches and on the boardwalk and spending their summers in beach bungalows. Some owned rental property and small businesses in Eastville (pseudonym). Eastvillers (pseudonym) are the non-White and White economically deprived, elderly, new immigrant, substance-dependent, "deviant," and formerly incarcerated population. They are the peninsula's excluded, exploited, and expropriated surplus population of the racial capitalist state who feel the brunt of inequities during disaster (Laster Pirtle 2020; Pulido 2016; Collard and Dempsey 2017).

I say that Eastvillers are the expropriated class because their bodies are the captured pawns through which State-funded capital flows to White economically privileged slum lords and White small business owners. Some of the Eastvillers to whom I spoke have been made unemployed and chronically economically deprived via economic exclusion from the 1980s with no opportunity to climb out of poverty. Already economically deprived urban residents are trapped in a system that on the surface seems to benefit them, when in reality it fills the coffers of those who use their bodies for profit. Similarly, developers are continually devising strategies to displace The Rockaways's economically deprived population, in order to grab the increasingly coveted coastal properties for their projects.

Slum lords and capitalists siphon transfer payments from the State through various programs designed to rehabilitate and provide the basic needs of The Rockaways's economically deprived. This racial capitalist, social services economy that has arisen in The Rockaways is the result of what many Rockaway residents see as

New York City discarding its social services–dependent populations in this coastal periphery, which is mainly out of sight. The capitalist, social services economy is visible from the neon "We Accept EBT [Electronic Benefits Transfer] Here" signs that flash from corner stores to the various forms of precarious housing paid with monthly government transfers. While they profit from governmental programs, the State does not hold accountable these slum lords who warehouse the impoverished masses in inhabitable conditions.

The racial capitalist state does not only facilitate the interests of for-profit corporations, developers, and landlords. Nongovernmental organizations (NGOs) lobby elected officials for contracts, public infrastructure investments, and favorable regulations (Marwell 2007). NGOs outside economically deprived neighborhoods are the recipients of this State funding (Marwell 2007). While NGOs have public-facing missions that deliver public goods, they connect to capitalists either directly or through their foundations. From the early 1900s, foundations served as tax shelters for multimillionaire capitalists and their families who wanted to evade estate or corporate taxes (INCITE 2007 cited in INCITE 2020).

This means that we need to scrutinize NGOs as we would for-profit corporations. We need to go beyond their missions and examine their institutional logics, practices, and work cultures. We also need to ask how the activities of NGOs in segregated urban communities may negatively impact their racially minoritized and economically deprived aid recipients. We especially need to examine whether and how they attempt to ensure equitable disbursements of public and government-subsidized resources.

Contrastingly, small community-based organizations serving in racially segregated urban areas step in to provide the economic and social safety nets that the State fails to provide. Community-based organizations benefit the increasing populations of Black and Latino residents (Marwell 2007) who have migrated to cities from the rural United States and the Global South. In the 1960s, African Americans', Puerto Ricans', and Mexican Americans' fight for economic and housing inclusion led to the first State-supported community-based organizations in New York City, which would serve as models for other cities (Marwell 2007).

Community-based organizations operate on a local neighborhood or neighborhood area scale with their missions typically oriented to providing services such as "affordable housing, childcare, drug treatment, cultural programs, services for the elderly ... [and] job training programs" as well as fighting homelessness, neighborhood self-revitalization, and legal advocacy for disenfranchised local residents (Marwell 2007, 4). Their volunteer and staff pool are typically local (Marwell 2007).

Local community-based organizations and small churches based in economically deprived and racially and socioeconomically segregated urban communities are the most attuned to the specific needs of the urban economically deprived but are also the ones that are often on the brink of extinction. These small, community-based organizations experience disconnection from foundation or State-funded capital. This unfortunately also means that they are severely underfunded despite their vital role in serving racially minoritized and economically deprived urban residents.

As we have seen, the United States, as a racial capitalist state, widens the gap between urban White economically privileged and non-White economically deprived communities. The State subjugates, exploits, and excludes the latter. Simultaneously, the State rewards White economically privileged citizens with freedoms, opportunities, and advantages, making it possible to live an entire life in oblivion. Furthermore, the State supports the interests of capitalists and organizations that disproportionately benefit White economically privileged citizens and their communities.

Contrastingly, small community-based organizations serving their majority Black, Latino, and economically deprived communities experience a disconnection from State resources. Therefore, in order to understand how disaster response reproduced race and class inequality after Superstorm Sandy, an urban disaster, we need to examine the role of the State through the Federal Emergency Management Agency (FEMA) and local government, nongovernmental organizations, and community-based organizations in the implementation of disaster response.

Superstorm Sandy: A "Post Katrina" Disaster

Superstorm Sandy made landfall in New York City on October 29, 2012, resulting in an urban disaster. Many of the city's essential and residential buildings lined the coastline, leaving over one million New Yorkers in the highest priority evacuation zone (Gibbs and Holloway 2013). Sandy's storm surges flooded lower levels of homes and apartment buildings and left eight hundred thousand residents without power, heat, or hot water for up to several months during cold temperatures (Gibbs and Holloway 2013). The sudden need to provide essential services and supplies led the city to coordinate a massive disaster response operation with large NGOs and local nonprofits to mobilize twelve thousand volunteers within the first eight months of the disaster (Gibbs and Holloway 2013).

Several studies have documented that New York City's racially minoritized and urban economically deprived disaster survivors shouldered the brunt of disaster inequality (see Faber 2015). These Sandy studies focused on how race and class shape disaster risk as well as vulnerability in coping with the storm (Faber 2015). These studies reveal that Black disaster survivors and the most economically deprived White disaster survivors, including the elderly, were more likely to live in flooded areas (Faber 2015). Black and Latino residents were also more likely to experience exposure to storm surges and flooding, because they tended to reside in public housing near coastal areas (Faber 2015).

Disruption to transportation impacted Black residents who already live further away from bus stops, constraining their access to networks, employment, and schools (Faber 2015). Sandy studies also emphasize how housing and the lack of health care shaped the vulnerability of the economically deprived urban residents during Sandy (Hernández et al. 2018). These studies also show that public housing residents, who were already economically, socially, and medically vulnerable, were without water, electricity, heat, and transportation for several weeks after Sandy. While these studies point to demographically associated disaster inequality, they leave us with the question of how race and class structures and processes infiltrate disaster response. This was a question I explored during my fieldwork in New York City's Brooklyn and The Rockaways, after Superstorm Sandy.

This book contributes to the archive of post-Katrina studies on race and class inequality in disasters. I cannot overstate the defining role Hurricane Katrina played in pointing disaster scholars to the centrality of race and class before, during, and after disaster. The case of Hurricane Katrina also established the significance of race and class inequality in the collective memory of Americans (Brunsma, Overfelt, and Picou 2007). The stark difference in the experiences of Black New Orleans residents from those of White residents, the disproportionate deaths among Black residents in the lower ninth ward, and finally the slow governmental response to the crisis led to public allegations of racism.

Prior to Katrina, discussions of race in the disaster literature focused on indicators such as race and socioeconomic status of individuals and measuring how these impacted risk and access to services (Bolin 2006). However, once it became clear that a lack of appropriate and timely governmental response became "the disaster" or "the crisis," a new wave of disaster research began to reconsider how we theorize the workings of race and class in racially minoritized and economically deprived urban areas.

We saw in New Orleans that decades of disinvestment by the local government in the infrastructure and welfare of the Black urban areas, and in particular the lower ninth ward, explain the "wider disparity in adaptation and recovery between Black and white storm victims" (Bullard 2009). We understand that Katrina is a man-made disaster that implicates the State in the loss of life and property of an already socioeconomically deprived, racially minoritized population. The case of New Orleans also revealed that the State and NGO response only served as a multiplier effect on longstanding race and class spatial inequalities (Bullard 2009).

A few years before Katrina, the Homeland Security Act of 2002 privatized emergency management and devolved the State's responsibility to private organizations such as corporations and large NGOs. The 2002 Act also reorganized FEMA as a purchaser and coordinator of services. The act also allows the federal government to contract out disaster recovery activities to private firms.

Rita J. King (2009, 169) writes that by the time Hurricane Katrina hit New Orleans and the rest of the Gulf Coast, FEMA was already "crippled by cutbacks and gutted of personnel." Neoliberal ideology reduced governmental intervention and emphasized calls for personal responsibility in disaster recovery. This ideological infiltration of disaster response has led to a shifting of the responsibility of response and recovery to local governments, communities, and citizens. States that were already experiencing fiscal problems would in turn rely on the private sector. Unsurprisingly, the move toward privatization of disaster response resulted in corporate interests playing a significant role in disbursing disaster aid, proposing redevelopment plans, and bidding on government contracts to rebuild disaster-impacted cities. This means that disaster response follows a market-oriented model of redevelopment that equates rebuilding communities with subsidizing business recovery and revitalizing financial centers (Gotham and Greenberg 2014).

Private interests often use crises as means to restructure under the guise of redevelopment. Naomi Klein (2008) theorizes the State's role in the relationship between disaster and capitalism: The State backs corporate interests. Corporations capitalize on the collective trauma that disasters unleash on the impacted population. They do this by opportunistically pushing through capitalist interests and simultaneously stripping preexisting social safety nets for vulnerable survivors. Kevin Fox Gotham and Miriam Greenberg (2014) argue that disaster redevelopment in New Orleans after Katrina and New York after 9/11 bolstered the French Quarter and made the Wall Street areas "vibrant and dynamic 24-hour communities," respectively, to the neglect of the lower ninth ward and Chinatown (45).

Katrina taught us to pay attention to the recovery divide across local geographies. Cutter et al. (2014) describe Mississippi's recovery after Hurricane Katrina as a "recovery divide" (13). They argue that the political elites, the business community, and the wealthy recovered the fastest, while Black and economically deprived disaster survivors suffered the greatest impact from both Hurricane Katrina and Camille, yet saw the slowest recovery. Katrina also taught us that State-funded disaster response prioritizes property over people. Most of the federal rebuilding funds allocated to Mississippi rebuilt federal buildings, coastal facilities, and ports, while the vulnerable populations were not able to access much of this funding (Cutter et al. 2014).

Katrina exposed how important racializing narratives around deservingness and undeservingness are to race and class inequality in the allocation of disaster aid. Department of Housing and Urban Development (HUD) funding disproportionately went to Mississippi versus New Orleans (Weber 2017). This was due to party politics and racial and class assumptions around Mississippi being deserving and New Orleans being undeserving (Weber 2017). Elected officials argued that New Orleans suffered an unnatural disaster due to a poorly maintained levee system. Therefore New Orleans's claims to recovery funds were unmerited. They simultaneously argued that Mississippi was deserving because its damages were due to the hurricane (Weber 2017).

Superstorm Sandy presents continuities to Katrina. Studying post-Sandy New York City also gave me the opportunity to gain an in-depth understanding of how race and class structures and processes intervene in disaster response. Through my fieldwork in Brooklyn and The Rockaways, with primary focus on Canarsie, Westville, and Eastville, I gained intimate knowledge and insight about what it means for urban, Black, Latino, Native American, and economically deprived White disaster survivors to have to navigate a complicated ecology of disaster response of unfamiliar nonlocal organizations, agencies, and responders.

Through this project, I have now personally seen the underbelly of disaster response at the person-to-person, block-by-block level. In the wake of disaster, it may seem that existing inequalities become less important, as everyone reels from the experience of acute crisis. However, I witnessed the persistence of race and class inequality and how it shaped disaster response. This is why I can no longer breathe that collective sigh of relief that we do after hearing on the news that a plethora of governmental and nongovernmental agencies have finally arrived in disaster areas. I now understand that when FEMA and nonlocal, large NGOs finally do get to economically deprived urban areas, far from disrupting long-standing race and class inequality, they help reproduce them.

Disaster response merely serves up old wine in new bottles. Old race and class logics of urban spaces combine with the color- and class-blind institutional logics of nonlocal organizations. These combined logics create an ecology of unequal networking opportunities that privileges the disaster resource capture of White middle-class and affluent disaster survivors and their communities. The emergence of this skewed ecology allowed White economically privileged Westvillers to outpace their Black, Latino, Native American, new immigrant, and White economically deprived neighbors in the adjacent impoverished community of Eastville.

Part of this disaster response ecology is the institutional logic of color and class blindness. These logics helped displace the services-dependent chronically economically deprived of Eastville. These logics also increased the bureaucratic burdens of Black, immigrant Canarsie disaster survivors and the racially diverse, economically deprived of The Rockaways. Furthermore, these logics excluded the undocumented immigrants, basement renters, and the self-employed disaster survivors with whom I spoke from recovering their disaster losses.

Ecologies of Inequity

INTRODUCTION

Superstorm Sandy

> Ethnographic discovery goes beyond the "sum of parts" of data collection and design. The researcher, who is the ethnographic instrument, must also adapt to unforeseeable developments that arise in the field. This entails continually making judgments regarding pursuing new paths of inquiry and observation. Discovery is the process and product of the ethnographer's intuition and insight. The incremental sharpening of insight equips the ethnographer to stitch together and keep alive the stories of communities.
>
> —Sancha Doxilly Medwinter, the author

Twelve days after Superstorm Sandy's landfall on October 29, 2012, I arrived in New York City from North Carolina with a truckload of donated supplies from a food and clothing drive that I, my husband, and my two daughters organized for those impacted by the disaster. From our living room, we had watched with millions of television viewers media coverage of the plight of disaster survivors. I thought even more about the possible plight of the many Black and brown faces of New Yorkers who we knew existed, but whose stories the news did not feature.

Prior to moving to North Carolina for graduate school, I lived in Brooklyn, New York, and graduated from one of the colleges of the city's public university system. Like many other Caribbean immigrants who are part of the city's salad bowl, I felt a sense of obligation to return to New York City and respond in some way. My family and I decided we needed to do something to help the disaster survivors. I sent out requests through my university department's listserv, while my girls, who were eleven and twelve years old at the time, created flyers with a list of needed donations and posted them on bulletin boards, doorways, and our apartment leasing office. I contacted a local radio station, which solicited their listeners for donations. My husband, who worked with a moving company at the time, put in a request for his employers to donate the use of a twenty-five-foot truck to transport disaster relief supplies. Within a few days, we had received enough donations to fill the truck that he would drive to New York City.

Prior to arriving in New York City, I wondered about the extent of the plight of these disaster survivors. I also knew that folks like Pastor Errance (pseudonyms are used for the names of all the clergy, staff, volunteers, and survivors mentioned in the book) would be on the front lines and that I would find my place wherever they needed me. I contacted Pastor Errance, whom I had known for years because of his mobile, bullhorn ministry that combined the gospel with disseminating immigration information to immigrant communities in Brooklyn. As expected, Pastor Errance had shifted his focus to bringing disaster relief to the same communities he regularly served. I asked him to take me to the communities that were impacted but had not received assistance. We went to Canarsie, a neighborhood in Brooklyn with a significant Black Caribbean and African presence.

We arrived at a street corner in Canarsie and opened up the back of the truck. Almost immediately survivors poured out onto the streets and surrounded the back of the truck. We began to disburse food, blankets, coats, and baby items. I also began my first set of interviews with Canarsie survivors who agreed to speak with me. Pastor Errance would later lead me to a local community-based organization in Canarsie where we donated the remaining clothing. Although Pastor Errance and other pastors were engaged in disaster response efforts, they did not have a large enough vehicle or a large enough space to store the supplies. In speaking with the volunteers, I also learned about the disaster response run by the Federal Emergency Management Agency (FEMA), which was set up in a large Catholic church on the other end of Canarsie. I was able to observe and interview disaster responders working at this site. The FEMA-run disaster response center also provided an opportunity to observe and interview disaster survivors.

In my conversations with Pastors Errance and Ward and Bishop Fabian, all Black West-Indian church leaders and now disaster responders, I also learned from them that The Rockaways was one of the hardest hit areas. They also talked about the difficulty of getting to The Rockaways since the storm had created an impasse. I wanted to see if that was still the case, and so the next day I asked a friend of mine to drive me to The Rockaways. Right before Sandy, she worked as a certified nursing assistant at one of the nursing homes lining the beach that was evacuated due to flooding. Public transportation had not yet fully resumed, and it was also during the winter months. I would subsequently have to make the trek on my own, which meant that keeping warm was always an important consideration. Our first visit to the peninsula consisted of driving around to survey the devastation and identify areas that were most visibly impacted by the disaster.

When I had decided to set out for New York City, my initial plan was to create a short video ethnography of less visible communities, which I would also circulate to disaster response organizations. I had envisioned conducting about thirty conversational interviews, in areas that we knew had been hit but had not received media coverage. This initial plan morphed into a much larger project of longer duration.

Over the course of my research, I interviewed a groundswell of 120 participants: disaster responders from FEMA, New York City, and New York State; nonlocal nongovernmental organizations; and local nonprofits, churches, and community-based organizations. I observed hundreds of disaster survivors and disaster responders in a variety of settings. Authoring this book allows me to present a synthesis of what I learned from those who are involved in the on-the-ground execution of disaster response. I also hope the book finds its way into the hands of undergraduate and graduate students with an interest in disasters and the reproduction of race and class inequality. Last but in no way least, I hope this book will reach the public, especially New Yorkers for whom I expect it will hold special significance.

Book Overview

In *Ecologies of Inequity*, I tell the story of how the economically deprived urban disaster survivors—a mixture of Black, Latino, Native American, White ethnic, new immigrant, public housing resident, homeless, and precariously housed—fared alongside their relatively more economically privileged homeowner neighbors. The story describes and explains the serial displacements of the service-dependent, chronically economically deprived by Sandy and the disaster logics of disaster response organizations. I also tell the story of how Black immigrant disaster survivors—undocumented, basement renters, working class, retired, and informally employed—fared alongside their Black middle-class homeowner neighbors in Canarsie. The story describes and explains Black immigrants' delayed assistance and hardships navigating the labyrinth of organization-mediated disaster assistance. The setting of this story is New York City, in the Canarsie disaster response area of Brooklyn and the Eastville (pseudonym) and Westville (pseudonym) disaster response areas in of The Rockaways (see figure 1).

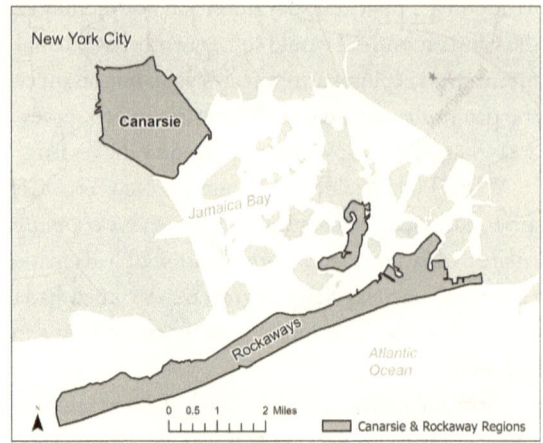

FIG. 1. Map of Canarsie and The Rockaways.

Disaster is an uncanny catalyst to observe in microscopic detail both the unraveling and reconstituting of features of urban environments and the uneven incorporation of social groups into urban spaces, which combine to accrue privilege to some urban residents while they relegate disadvantage to others. The dynamism of emergent disaster response areas unearths cumulative mechanisms of privilege and disadvantage that ordinarily stitch and dissolve into the familiar fabric of urban life, which is often too subtle, ubiquitous, and slow-moving to uncover their production.

The disaster response provided by governmental and nongovernmental organizations is a critical opportunity for the possible recalibration of long-standing race and class inequalities embedded in urban spaces. However, governmental and nongovernmental organizations miss this opportunity entirely. Race and class processes intervene in official disaster response. This occurs from the earliest point of contact, the placement of disaster response centers, to the cognitively mapped-out spatial perimeters of disaster response among volunteers, to the emergence of diverse types of responder-survivor expectations, relations, and information sharing.

The entry of nonlocal entities such as FEMA and NGOs marks the reorganization of disaster areas into a new *ecology of inequity*, reconstituting White privilege and urban economic deprivation. This emergence of an *ecology of inequity*, which I explain in detail in the next chapter, escapes the purview of citizens as this occurs at a moment when most are still focused on the spectacle of physical destruction, fatality statistics, and the altruistic acts of first responders.

Chapter Outlines

This book, in eight chapters, successively tackles the empirical puzzle: Why and how do the racially minoritized and economically deprived urban disaster survivors lose ground to their White economically privileged neighbors after disaster? What is the role of disaster response in the reproduction of race and class inequality?

Chapter 1 reformulates this puzzle of disaster inequality as a question of nested processes of racializing, classing, and social capital (networking) relational stratification. This framework sets the stage for an investigation into how and why urban spaces, organizational environments, and social group hierarchies and relations may combine and operate to undermine the egalitarian distribution of public goods. In order to answer this question, the chapter continues the discussion of the concept of *ecologies of inequality* that I introduced in the introduction.

I argue in the chapter that ecologies of inequality create an *ecology of privilege* and an *ecology of deprivation*. These opposite sides of the same coin deliver the favorable and unfavorable spatial alignment of networks of governmental and nongovernmental organizations and their responders across urban areas differentiated by race and class. *Ecological Privilege* provides sustained, instrumental benefits for already privileged groups. The distributional impact of this ecological privilege translates into ecological deprivation for adjacent areas inhabited by racially minoritized and other marginalized groups.

Chapter 2 begins with a description of the racial and class topography of The Rockaways. I draw on observations and conversations with Rockaway disaster survivors and local disaster responders with tremendous local knowledge of their communities as well as my observations. The chapter then goes on to describe how these racial and class logics and topography of The Rockaways seep into the disaster response process from the placement of centers to the spatial perimeters set by disaster response site managers and volunteers. The chapter also discusses how the perceptions of disaster responders translate into racialized, classed zones of disaster response.

Chapter 3 similarly describes the racial and class topography of Canarsie, Brooklyn. I also explain the segmented incorporation of immigrants into the United States and how the refraction of legal or undocumented status extends to the segmentation in the receipt of FEMA disaster assistance. These statuses of citizenship intersect with homeowner and renter, specifically basement renter, statuses. My dialogues with Ferdinand, the nonlocal FEMA disaster response site manager, and his staff and volunteers from a variety of New York City and State agencies captured the hardships of Black immigrants. I present what I learned from these responders, Canarsie disaster survivors' challenges with securing disaster assistance.

Chapter 4 opens with an introduction to Canarsie women who are employed Black working-class mothers. These women have become narrators of their own stories of hardship. The women describe their experiences of destruction and dispossession of their personal capital. They express their belief that they were forgotten by FEMA and the local government. They are frustrated that they are left to fend for themselves and their children. I discuss the importance of Black and economically deprived women's informal networks to their survival in the immediate aftermath of disaster. The chapter then moves into the on-the-ground reality of the difficulty with which Canarsie disaster survivors navigate FEMA's labyrinthine grant appeals process, which reroutes them to insurance companies and SBA.

Chapter 5 answers the question of what happens to social capital during crises and disasters. I illustrate how a disaster event erodes or suspends social network capital through social tie fragmentation, depletion, and deflation. I illustrate the similarities and the differences for White, economically privileged urban disaster survivors and survivors who are racially minoritized and economically deprived.

I introduce a newly coined concept, *crisis capital*, a transient form of *relational capital* that emerges in disaster areas. Crisis capital provides a much-needed support when the viability of mature networks is compromised. However, this crisis capital is short-lived and insufficient. Therefore, communities need this crisis capital supplemented with a more sustainable form of relational capital. This supplementation occurs in Westville, the White economically privileged residential area, through the amicable relations with a high resource, nonlocal NGO. However, the same NGO seemed to merely "take over" in the racially and ethnically diverse, economically deprived disaster area. This move disrupts the crisis capital of this community.

Chapter 6 explains how the on-the-ground operations of disaster response centers run by FEMA and NGO managers usher in a *logic of response* that replaces the *logic of services*, effectively displacing the service-dependent and economically deprived disaster survivors in Eastville and the Canarsie area. This disaster response logic weaves middle-class bias and color and class blindness—through norms, expectations, practices, and assumptions—into disaster response. Together, these complicate the ability of the most marginalized and invisible to access resources. I present my observations and conversations with disaster survivors who had visited FEMA, New York State (NYS), and NGO disaster response centers. I also present interviews with FEMA, NYS, and the leader of a large church. The chapter ends with Richy's story, which demonstrates how this displacement of the Logic of Services by the Logic of Response lets the chronically economically deprived slip through the cracks.

Chapter 7 conceptualizes social capital as an interactional privilege, an important aspect of *Ecological Privilege*. White economically privileged Westvillers are able to create new instrumental social capital. Eastville's ecological deprivation bars them from these interactional opportunities for creating new social capital with NGO responders.

Chapter 8 presents a flashback of how ecologies of inequity emerged in Canarsie and The Rockaways. The chapter presents four types of disaster response organization networks that emerged in Canarsie and The Rockaways: *organization agglomeration, organization isolation, organization hosting*, and *organization coalition*. These different configurations of organization networks emerged out of diverse types of institutional and spatial urban contexts. However, all involve FEMA, large and small NGOs, large and small local churches, and community-based organizations, but with varying relations. The types of relationships that connect these organizations help explain some of the nuances in the ways that Sandy survivors in Canarsie and The Rockaways experienced organization-mediated disaster response.

In the conclusion I provide a brief overview of the findings of all the data chapters of this book. I explain the role of governmental and nongovernmental organizations in creating an ecology of inequity. I make the argument for making race and class central to policies and practices of governmental and nongovernmental organizations. On the policy front, I provide a critique of the November 2020 National Advisory Council's 2045 policy reform agenda for FEMA. This report only makes a cursory mention of race in discussing equity in disaster response. On the practice front, I make recommendations to nonlocal NGOs about how to avoid creating ecologies of inequity.

The epilogue is a casual conversation with Bishop Fabian and Freddie, the founder of Always With You, about the current state of post-Sandy Canarsie and Eastville. I present their reflections on the political, institutional, and organizational field since Sandy and how these continue to shape the financial and social capital of local churches and local community-based organizations serving the urban economically deprived, particularly within our current Covid-19 reality.

Appendix A, my interview guide, follows the epilogue. In Appendix B, I discuss my positionality in the field, my encounters with racial prejudice, and my suggestions for how to ethically and empathetically navigate the study of human beings experiencing abject circumstances, as a researcher. I end with a discussion of the strengths and limitations of the study. Appemdix C presents demographic information on Canarsie and The Rockaways.

Reading This Book

In writing this book, I use pseudonyms such as "Westville," "Eastville," "Resiliency is Us," and "Always With You," to protect the privacy of research participants relaying personal experiences. I have also extended the use of pseudonyms to clergy and disaster field site managers who were providing expert explanations of the on-the-ground organizational practices. It should be noted that most participants asked that I use their real names. Beyond these pseudonyms, I also take additional steps to obscure locations. While the book provides demographic tables, graphs, and cartographic maps for Canarsie and the Rockaway peninsula, following recommendations of reviewers to provide an anchor for readers, these should be read as "in the general vicinity of" the sites of data collection. In discussions of parameters of neighborhoods and urban disaster areas, I rely on cognitive maps, in lieu of cartographic maps, described by research participants. I also frequently use "The Canarsie area" and "The Rockaways" to signify a less rigid treatment of these neighborhood boundaries.

I have taken additional steps to further obscure sites. I follow some other naming conventions that need clarification. Throughout this book, I use adjectives such as "FEMA-run" or "NYS-run," and "NGO-run" centers, to remind the reader of the de facto relative, operational dominance of managers of various types of organizations at a particular center. It is important to note, however, that FEMA was present only on invitation to assist with the local efforts of disaster response. Furthermore, the centers themselves are non-permanent assemblages of relief organizations, governmental and nongovernmental, providing invaluable disaster assistance to disaster survivors. The venue of these disaster response centers were primarily provided by large churches already serving these communities. My use of "disaster response center" also encompasses "relief" "recovery" or "restoration" centers.

I also provide some guidance on reading the narratives in this book. The chapters shed light on the challenges of navigating organization-mediated disaster assistance. These narratives center the stories, reflections, and perspectives of those impacted by disaster and disaster assistance.

The narratives of these disaster survivors reflect how impactful the challenges they experienced were, and how they were making sense of their own experiences. Drawing on the tradition of Black Feminist methods, I select words such as "reflection" on experiences versus "perception" of experiences, as I argue that the latter may perpetuate the default invalidation of recounted harm experienced by members of racially minoritized people. As such, my role is not to interrogate the expressed understandings and experiences of my research participants. However, I do also present in this book, the explanations offered by disaster responders of standard organizational practice, on how programs are designed to work, and how official information and processes are meant to be understood and navigated.

Alternating among these multiple perspectives and experiential knowledges, and awareness, I illuminated the difference between intent and experience. This approach makes legible the inadvertent challenges and consequent harms emanating from the on-the-ground execution of disaster assistance. Finally, it is worth reminding the reader that the goal of the narrative presentation in this book does not seek to generalize a population, evaluate a program, nor does it claim that the experiences depicted were typical among disaster survivors in the disaster areas in the study. Instead, the book's narrative helps us gain an in-depth understanding of a process as it unfolds in the lives of my research participants in their specific meaning context.

The Ethnography

The primary field sites for my ethnography, what I refer to as disaster response centers (i.e., relief, recovery, restoration), and their surrounding vicinities in Brooklyn and The Rockaways. I conducted expert interviews with the field site managers, who were from FEMA, New York State, a large, nonlocal NGO, and one community-based organization. I also interviewed leaders of local churches who rendered disaster assistance in Brooklyn and The Rockaways. I povide some demographic and geographic descriptions and maps for Canarsie and the Rockaway peninsula to help anchor the ethnography for readers. However, I demarcate urban disaster areas and neighborhoods according to the cognitive maps and the notions of belonging of research participants. Locational details of interviews and observations do not rigidly follow cartographic boundaries. For example, my use of "Canarsie" should be understood more broadly as in the vicinity of "The Canarsie area" or serving Canarsie disaster survivors. Similarly, in discussing an organizational practice that is corroborated across field site managers, I transpose some of the quotes of field site managers. I also use pseudonyms for NGOs and surrounding disaster response areas as "Westville" and "Eastville."

The disaster survivors and responders with whom I spoke were primarily from the Canarsie area, which I refer to as Canarsie, and The Rockaways. In Canarsie, most of my interviews were with Caribbean immigrants, but I also learned of the experiences of African immigrants from the disaster responders at the FEMA-run

Canarsie disaster response center, volunteers at a local community-based organization, and church leaders providing disaster response in Brooklyn. In The Rockaways, I conducted interviews with Irish White disaster survivors in Westville and non-Irish White disaster survivors, who tended to reside near the Eastville-Westville boundary. The latter were working-class renters, retirees, and immigrants from European countries such as Poland, Russia, and Italy. I also interviewed several Native Americans. Among the disaster survivors, who were phenotypically Black, some identified as American citizens with no migration in their family history, and others identified as part of the Caribbean diaspora.

Demographics and Geography of Participants

The frigid winter winds and disrupted transportation routes within the first few weeks meant that I mostly captured the experiences and perspectives of able-bodied residents who could walk to these centers. I alternated between canvassing and observing disaster areas, where I spoke with disaster survivors while they were either standing in front of their homes or peeking outside from their foyers. Other disaster survivors graciously invited me in to see their waterlogged basements and debris-filled backyards. I also spoke with interviewees sitting at tables in response centers, doing laundry at the laundromat, walking down the street, waiting at the bus stop, and even riding the bus. Other interviews took place in churches, at community meetings, and in business establishments.

Most people I approached agreed to talk with me. Only four declined my request for an interview. I attribute this willingness to participate partly because I was pursuing a broader social goal of making my findings available to disaster managers and others who are in a position to improve the efficacy of disaster response and consequently the experiences of disaster survivors.

Among disaster survivors whose activity I could observe, my primary recruitment goal was to capture nuanced perspectives on important dimensions relevant to theory, such as race-ethnicity, employment status, housing type and tenure (Adeola and Picou 2012); and migration or moving experience (Donner and Rodriguez 2008). As I gained more insight into prevalent economic circumstances, I inquired about transfer receipt through social services. Conversations with responders and disaster survivors often provided the source of further observations and inquiries within a particular disaster response center and surrounding disaster response areas. These included residential clusters and other sites of which I may not have been previously aware but that would later prove crucial to my central question of uncovering specific mechanisms of disaster response inequity.

My data collection was necessarily multimodal, including in-depth unstructured interviews, semi-structured questionnaires, and observation, both participant and direct (O'Leary 2005). I adapted interviews to suit participant preferences, time constraints, and comfort, alternating among video and audio recording, pen and paper recording during an interview, or informal talk without the formality of note-taking. In an incremental way, each observation and interview allowed me to sketch the contours of working theories. These theoretical sketches did evolve and eventually coalesce as I gained more data, intuition, and insight from time in the field. I achieved rapport with key research participants and had the opportunity to continually reflect on and confirm what I was learning about salient aspects of the meaning context.

The study participants relayed to me their disaster experiences, the complexity of navigating the terrain of organization-mediated disaster response while they were still experiencing these hardships of disaster. This means that I also had the opportunity to triangulate, through observation, some of the experiences of which they spoke even as these unfolded over the course of several months (Small 2009a; Burawoy 2003). The experience of disaster and the process of seeking disaster assistance do not merely involve the individual, but the network of relations that include the immediate and extended family and friendship networks, and neighbors of the disaster survivor. Sometimes while I interviewed a participant, the spouse or partner would later join in. I kept these interviews linked so that I did not lose meaningful context. Many people who were there alone volunteered information about their partners, despite their absence shared the disaster experiences of their partners, including where these experiences differed.

While I went to the field equipped with an interview guide and questionnaire (see Appendix A), most interviews ended up being mostly organic and conversational. Beyond the early response period, I was able to follow the interview guide more closely, either before or following organic conversations. My initial interviews felt scripted, awkward, and inappropriate for a post-disaster setting. My white paper with my questions served as a "white coat" that only separated me from disaster survivors. I began to feel like I was interrupting the flow of what these disaster survivors really wanted to share about their disaster experience. In fact, after answering all my scripted interview questions with quite terse responses, one disaster survivor at the laundromat asked who prepared these questions. She critiqued the more detailed, structured questions. I had included the position generator instrument, a measure of social capital used in survey research that asked about their social connections, the occupations of people they knew, and who offered to help.

In a playful tone I said with a smile, "Well, that would be me," at which we both laughed. It was only then that she became more relaxed and began to tell me about her neighbor who was displaced from her basement and that no one had heard from her. She expressed how much she actually wanted to help, but because her neighbor had gone to a shelter, she could not reach the woman and had no way of helping her. The conversation proved most illuminating for the rest of my research, as I was more attuned to the plight of basement renters.

This exchange with this disaster survivor also began my future inquiry into the utility of pre-disaster social capital during moments of crisis. At that moment I decided that a more organic exchange would yield more interesting insights and make the interviews more pleasant for my research participants, who were already enduring so much. However, progressively moving beyond the early response period allowed me to rely on the interview guide more closely to capture specific details, either at the beginning or end of more organic conversations. I decided to approach my conversations with disaster response site managers with less formality. My frank conversational style gave me an opportunity to engage disaster responders in thinking more critically about the interactional context of the disaster response centers that they were managing and working in. I am hopeful that our exchanges also presented reflexive considerations of more equitable delivery of disaster assistance.

Ethnographic Discovery

Ethnographic discovery goes beyond the "sum of parts" of data collection and design. The researcher, who is the ethnographic instrument, must also adapt to unforeseeable developments that arise in the field. This entails continually making judgments regarding pursuing new paths of inquiry and observation to fill gaps made visible *only* through iterative, reflective assessment of most recently attained data vis-à-vis previously collected data. This iteration of reflection-propagated return to the field relies on the ethnographer's intuition and insight. Both of these skills sharpen as part of the product of ethnographic discovery. Over the duration of the ethnography, this incremental sharpening of insight equipped me to pull together and construct the narratives of communities surviving disaster. I weaved this metanarrative out of the many "stories" I learned during conversational interviews and observations.

Many of these conversations with disaster survivors and responders, whom I sat with, followed, and assisted, occurred primarily during the first six months, the height of organization disaster response. Subsequently, I would return, over the course of two years, for fourteen visits of two-to-four-day stints of twelve-to-fourteen-hour shifts, to these disaster response areas. During my returns to the

Introduction

field, I was able to recap and follow up with key informants and some research participants from the earlier response efforts. These were follow-up in-person conversations with disaster survivors, community-based organizations, local churches, and small business owners who had reopened their establishments.

Ethnographic Analysis

A central focus of my analysis included identifying, describing, interpreting, and theorizing the similarities and differences in interactions and informational exchanges among disaster survivors and responders. Comparison and triangulation were centerpieces of my analysis of observational and interview data. I distinguished what volunteers and site managers said about the assistance they offered disaster survivors from my own observations of these interactions and assistance with disaster survivors (Blumer 1958). I also triangulated what site managers said with what volunteers said about the quality of services provided. In some instances, I followed up on particular observations where a disaster survivor may not have received a particular kind of assistance, in order to capture their own assessment of their needs and whether they thought these needs had actually been met by responders. In order to accomplish differential experience, I compared these relations across locally meaningful distinctions in racial, ethnic, immigrant status, class, and urban areas.

Specifically, my analysis also entails comparing and explaining the interplay among urban spaces and organizational and networking/relational environments. I paid attention to how these dynamics intervene in social interactions on display in plain sight during disaster response. For this part of the analysis, I took seriously the intersubjective meanings (Blumer 1958) the participants made of themselves and their relation to others in the local environment in order to explain the social distance and symbolic boundaries among residents of these communities (Bourdieu and Passeron 1977; Lamont and Fournier 1992).

As earlier observations and interviews provided leads, I used what I incrementally learned to triangulate among the meanings (Blumer 1958) volunteers and responders assigned to their own observations and interactions with disaster survivors. Finally, I present my results that capture the "main story underlying the analysis" (LaRossa 2005, 850).

During my analytical process I was able to capture the racial and class implications of the on-the-ground organization and operation of disaster response. I also connected these to the changes in the ecological environment of the disaster areas I studied (Klinenberg 2002). Traveling back and forth to different disaster response areas, sometimes within the same day, allowed me to gain deeper insight into points of convergence and divergence in the collective experiential and meaningful contexts of disaster response across various urban landscapes.

The multilevel, multisite design of my ethnographic study yielded uncommon opportunities for fine-grain comparisons. I was able to comparatively analyze the experiences and accounts of Eastville, Westville, and Canarsie disaster survivors with the same NGO, Resiliency Is Us. Another opportunity for comparison was having the same site manager for both the Westville and Eastville NGO centers. Yet another was being able to compare the relationship that Resiliency Is Us established with both Westville and Eastville upon entry. I was fortunate to have met two local informants who began volunteering at the respective sites, even before the nonlocal NGO arrived. I also gained tremendous insight from my interviews with a handful of Eastville disaster survivors in Eastville, whom I quite by chance had an opportunity to observe and interview again at the Westville center.

I had similar comparative opportunities with disaster response centers run by New York State and by FEMA. I was able to compare disaster response approaches at two FEMA sites in The Rockaways (outside of Westville and Eastville) and Canarsie. I compared management styles of site managers at New York State and FEMA disaster response centers that offer the same suite of services. I was able to compare local experiences of leaders of small storefront churches engaged in disaster response with those of leaders of large churches, both within and across Brooklyn and The Rockaways.

Altogether, the multilevel design of this study allowed me to "see" the structures and processes of disaster response. The ethnographic design allowed me to trace how disaster response uniquely shaped the opportunities and constraints of disaster survivors over time. The comparative design across many contexts enabled me to capture crucial differences across race- and class-differentiated communities. The unique opportunity to have captured the disaster response experiences of Sandy survivors, told from multiple perspectives, has charged me with the responsibility of crafting an empathetic, nuanced synthesis of what I was fortunate to learn.

CHAPTER 1

Ecologies of Inequity

> This ecology emerges due to destruction of structures and the placement of makeshift disaster response centers that reorganize mobility and the use of space. Similarly, the influx of thousands of volunteers and the dislodging of disaster survivors from their residences structure the frequency, tenor, and content of relations. Networks of family, friends, and churches reconfigure and change functions as they too are impacted by disruption in transportation and destruction of infrastructure and buildings.
>
> —Sancha Doxilly Medwinter, the author

We know from disaster research that non-White and economically deprived disaster survivors are far less likely to recover losses from disaster than those who are White and economically privileged (see Fothergill and Peek 2004). We can attribute this lower prospect for recovery among those who are already socially and economically deprived to their being at the highest economic and psychological risk (Adeola and Picou 2012; Fothergill and Peek 2004; Cutter, Mitchell, and Scott 2000). The disaster literature attributes these stark disaster disparities to the social vulnerability of these populations.

William Julius Wilson (2012) argues that the urban economically deprived Black population is spatially segregated from both White and Black middle classes, leaving them in a condition of concentrated disadvantage. The racially minoritized and new immigrants concentrated in urban areas become socially vulnerable because of segregated opportunities, along the axes of race and class, in housing, employment, and financial capital (Donner and Rodriguez 2008; Oliver and Shapiro 1995; Massey and Denton 1993).

While pre-disaster inequities create an uneven playing field going into disaster, we also need to examine pathways of inequity during the disaster response period. We know from the disaster literature that the racially minoritized and the economically deprived receive less institutional aid (Bates 1982; Drabek and Key 1984). The execution of official disaster response amplifies long-standing race and class inequalities among disaster survivors (Barnshaw 2005; Oliver-Smith 1986; Peacock, Gladwin, and Morrow 1997).

In order to uncover the inequitable pathways to race and class inequality during disaster response, we need to make visible structures and processes that perpetuate race and class inequality before disaster. (I present a lengthier explanation of why the United States is a racial capitalist state in the prologue.) The first step in this process is recognizing that race is baked into the foundational structures and institutions that continue to exist today. Michael Omi and Howard Winant (1986) argue that in the United States, "every state institution is a racial institution" (83). This means that even state organizations and laws that manage disaster response "allocate differential economic, political, social, and even psychological rewards to groups along racial lines" (Bonilla-Silva 1997, 474). This means that the institutional logic of the racial (capitalist) state also permeates the practices of well-meaning disaster response organizations and volunteers (Hoelscher 2003; Omi and Winant 1994). Furthermore, institutional logics that permeate urban space and organizations and that animate community relations play a significant role in manufacturing inequality during disaster response.

Ecology of Inequity as Nested Structures and Processes

I argue that recognizing that an *ecology of inequity* emerges during disaster response is important to our understanding of how disaster response creates disaster inequality. First, I explain what I mean by the term "ecology." My specific articulation of "ecology" builds off of Eric Klinenberg's (2015) conceptualization in his book *Heatwave*. Klineberg (2015) invokes the concept of ecology in describing North Lawndale, where elderly African Americans disproportionately died of heat-related deaths. According to Klinenberg, North Lawndale was a "dangerous ecology of abandoned buildings, open spaces, commercial depletion, violent crime, degraded infrastructure, low population density, and family dispersion" (91).

Klinenberg (2015) illustrates that the lack of informal networks of family, friends, watchful neighbors, and churches and the lack of service and institutional support led to the racial disparity in rates of heat-related death from 100-degree temperatures in the 1995 Chicago heatwave. My conceptualization of ecology similarly captures the spatialized social network and social capital dynamics.

Unlike Klinenberg's ecology, the ecology of spatialized networks of organizational and interpersonal relations that I describe is an emergent one. This ecology emerges due to destruction of structures and the placement of makeshift disaster response centers that reorganize mobility and the use of space. Similarly, the influx of thousands of volunteers and the dislodging of disaster survivors from their residences structure the frequency, tenor, and content of relations. Networks of family, friends, and churches reconfigure and change functions as they too are impacted by destruction in transportation, infrastructure, and buildings.

Next I explain what I mean by the term "inequity." My emphasis on the inequity aspect of this emergent ecology points to the cumulative privilege and abundance stemming from the presence of a disaster response machinery in one disaster area, and how this necessarily means cumulative socioeconomic resource disadvantage and deprivation to another disaster area from its absence. One aspect of privilege is spatial privilege. In understanding the racial significance of this spatial privilege, I draw on Pulido's (2000) conceptualization of White privilege as a "structural and spatial form of racism" (12).

Similarly, my emphasis on socioeconomic resource deprivation relies on Galtung's (1969) conceptualization of structural violence, which emphasizes the psychological and physical suffering meted out by unjust institutions. Galtung (1969) emphasizes that these conditions that impact marginalized segments of society are not only unjust but preventable because of the current institutional capacity and technological advancement. Furthermore, the needed resources are available, as they are already enjoyed by the more privileged groups in society (Galtung 1969).

Spatial inequality has also been a focus in the fields of critical environmental justice and disaster studies. These point to the race and class spatial inequality in exposure to environmental risk and timely receipt of disaster response. Black, Latino, and Native American economically deprived residential areas, when compared to adjacent majority White areas, expose these inequalities across space (United Church of Christ 1987; Bullard 1983; Brunsma, Overfelt, and Picou 2007; Bullard and Wright 2012).

Segregationist policies in response to the Great Migration of approximately two million African Americans between 1910 and 1950 from the South has left a blueprint for the unequal development of northern cities such as New York City (Taylor 2014). Parting with the vantage point of most environmental justice scholarship, Laura Pulido (2000) reverses the emphasis on the disproportionate pollution in majority Black areas. Instead, she asks: "How did Whites distance them-

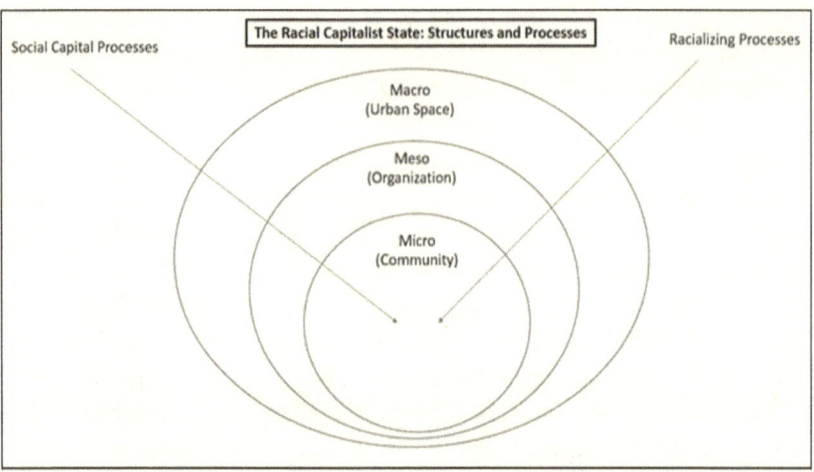

FIG. 2. Disaster inequality as nested structures and processes.

selves from both industrial pollution and non-Whites?" (2000, 14). I ask and answer a similar question in my study in The Rockaways: How does a White economically privileged community help exclude an adjacent racially and ethnically diverse, economically deprived community from pertinent disaster information and resources during disaster response?

Ecologies of inequity, the spatialized networks of urban spaces, organizations, managers, volunteers, and staff, shape the chances of receiving timely access to disaster resources. These disaster response ecologies configure relational inequities that span the nested structures. In order to trace the emergence of ecologies of inequity, we need to simultaneously examine both networking and racializing processes across the levels of urban spaces, organizations, and communities (see figure 2). In the following section I describe how each of these levels contributes to the formation of ecologies of inequity.

Racializing Urban Spaces

At the macro level, the boundaries inscribed in urban space make visible the racial and class hierarchy in the local context. This can be in the form of physical structures and infrastructure that restrict residency and movement of certain classes of residents (Blumer 1958; Bobo and Hutchings 1996). We see how racial hierarchy can be inscribed in space, by understanding that notions of racial difference are tied to a *sense of group position* relative to others within a specific context (Blumer 1958; Bobo and Hutchings 1996). For these reasons, I capture the rank ordering of the local racial, ethnic, and class groups through the accounts of The Rockaway and Canarsie disaster survivors and responders.

Urban space also sets the stage upon which social categories of race and ethnicity are contested (Barth 1969; Lieberson 2000). Jonathan Rieder's (1985) study of the territorializing of New York City informs my understanding of how urban space animates ethnic group relations (Sidanius and Pratto 1999). An analysis of the territorial nature of urban space is crucial for understanding the power relations, influence, and privilege wielded by community members within and across their residential clusters.

Racializing Organizations

Connecting the macro level to the meso-level, the race and class topography of urban spaces help structure the distribution (Gotham and Greenberg 2014) of disaster response organizations across disaster areas. Disaster response organizations are important actors in creating ecologies of inequity. Organizations are both racialized and racializing. On the first point, I draw on Victor Ray's (2019) conceptualization of organizations as "meso-level racial structures" (31). They legitimate resource inequality across White and non-White organizations (27). Therefore, rather than assuming race neutrality, we should understand organizations as being White or non-White.

According to Ray (2019), White organizations are institutions that benefit from institutional support of White dominance. White organizations have large endowments of private resources that far outnumber the resources of Black and non-White organizations (Ray 2019; Wooten 2015; Frazier 1957). Organizational segregation also ensures that while Black people patronize White organizations in White spaces, the reverse is seldom true (Ray 2019; Baradaran 2017). On the second point, as racializing conduits, organizations accept, transform, and transmit external racializing logics of institutional fields and urban places and spaces.

I draw on an organizational capital inequality approach (Small 2009b) in the ethnographic design of this study. I also draw on Nan Lin's (2001) conceptualization of social capital inequality as resulting from relational ties with contacts whose social capital stems from their unequal positioning within a status hierarchy. I integrate these perspectives to arrive at a conceptual framework that allows me to understand disaster response centers as high or low resource sites of interaction, information, and resource exchange. These are capable of forming social network/social capital ties both among embedded actors and among organizations themselves.

A more recent publication by Donald Tomaskovic-Devey and Dustin Robert

Avent-Holt (2019) confirms the importance of this relational interpretive framework for explaining how organizations help broker inequality. I further conceive of organizational sites as forming network ties with urban spaces and their communities. Drawing on Melissa Wooten and Andrew J. Hoffman's (2016) theorizing of organizational fields, Daniel Bolger (2021) also argues that these organizations follow an institutional logic of place and space that orients toward their stakeholders within an organizational field, not just service recipients.

Bolger (2021), in his study of two faith-based organizations in Houston, Texas, finds that these neighborhood service organizations assign their target service community a "Brown" signifier to the exclusion of "Black" racial signifiers. This designation mimics the "safe" versus "unsafe" understanding of spaces (Bolger 2021; Gotham and Brumley 2002). Such designations allow these organizations to attract White donors and volunteers who live in nearby wealthier areas (Bolger 2021). I also draw on Ray's (2019) discussion of racial segregation as a cultural schema, made real in its ability to restrict organizational resources from marginalized groups. These insights on organizations help explain why certain ecologies emerge in some urban disaster areas and not in others.

Racializing Networking Environments

Moving from the meso-level to the micro level, the racialized and classed access to organizations leads to unequal opportunities for networks and networking among disaster survivors and responders. Informal networks play a significant role in accessing disaster assistance. Social networks and their attendant resources, social capital, are crucial to bouncing back from the loss and harm associated with disasters (Picou, Marshal and Gill 2004; Aldrich 2012).

We know that during disasters informal networks link disaster survivors to housing, disaster aid, and psychological support (Dynes 2002; Fritz 1961). We also know that social networks help reduce inequities in the distribution of disaster aid (Barnshaw and Trainor 2010; Ritchie 2004; Peacock et al. 1997; Pelling 2003; Klinenberg 2004; Litt, Skinner, and Robinson 2012; Drabek and Key 1984). This means that if we want to make visible the pathways to disaster inequality, we need to analyze the loss, survival, and creation of social networks during disasters and disaster response.

Drawing on lessons from Hurricane Katrina, we see that the informal networks of Black women were an indispensable source of social support during the disaster. During the initial stages of Katrina, pre-disaster mutual aid among Black women morphed into evacuation networks that helped women flee the impending harm before Katrina (Litt 2012). Women provided and relied on "women-centered networks of care" that served as a source of emotional support (Weber and Peek 2012, 167). These social networks were especially important for economically deprived women who did not have the resources of middle-class women.

Pre-disaster networks, albeit vital, become fragile as a result of the disaster event. Research also shows that evacuation, long-term displacement, and depletion of needed physical resources erode the social networks of disaster survivors (Litt, Skinner, and Robinson 2012; Litt 2012; Elliott et al. 2010; Barnshaw and Trainor 2010; Barnshaw 2005: Peacock et al. 1997; Pelling 2003). For instance, Elliott et al. (2010) found that displacement compromised the capacity of disadvantaged lower ninth ward residents to "tap translocal ties." Because we know that the State's execution of evacuations disproportionately disrupted pre-disaster networks of urban economically deprived Black survivors, it is crucial that we examine the connection between governmental organizations and the social networks of survivors. We especially need to pay attention to the possible impact on the social networks of the racially minoritized and the economically deprived.

Social networks are important to disaster recovery, largely because they bridge the gap between disaster survivors and organization mediated disaster resources. In other words, social networks provide access to social capital. Social capital theory considers the role of familial and friendship ties and contacts as yielding expressive and psychosocial benefits, contrasted with the instrumental and material benefits linked by an NGO responder (Lin, Woelfel, and Light 1985; Granovetter 1973).

I adopt a network theory of social capital, which emphasizes the information and resources informally accessed and mobilized by individuals by virtue of their social ties (Lin 2001). A social capital-yielding tie is an interpersonal connection with an actor linked to resources through wealth, status, or high resource organizations (Lin 2001; Small 2009a). The social capital value of this connection hinges on conferring information or resources to a tie without a direct connection to such resources (Lin 2001; Bourdieu 1986).

In order to fully interpret the micro-level interpersonal relations, I also rely on the work of social psychologists who have pointed to less tangible aspects of interaction. Emotional and affective displays, meanings, and attachment in dyadic, group, and community relations mediate interactions (Lawler, Thye, and Yoon 2009). This perspective allows me to move beyond merely a transactional view of disaster assistance and examine the informational, resource, and affective exchanges among survivors and responders, who are the custodians of disaster resources in the context of disaster response.

My approach to social capital as a networking process, rather than a static characteristic of social ties in a social network, gives me leverage in ethnographically uncovering its role in the process of inequality. The status attainment branch of social capital theory has focused exclusively on quantitative studies of the utility of social capital in securing socioeconomic advantages (Lin, Ensel, and Vaughn 1981; Coleman 1990; Granovetter 1973). Research studying the utility of network social capital has also primarily focused on the role of high-status contacts on the outcomes of low-status job seekers in a competitive labor market. These studies tend to highlight the importance of informal job attainment strategies (Lin, Fu and Hsung 2001). I extend this conceptualization of social capital to seeking disaster assistance in an environment of scarcity.

Divergence in the process of social (network) capital across survivors and communities sheds light on the pathway to disaster inequality. Other relational processes may interrupt the relational process of social capital. Charles Tilly's theory of durable inequalities (1999) informs my relational analysis of the reproduction of race and class inequality. According to Tilly (1999), beliefs, symbols, and practices are central to unequal distribution of resources across social groups. I extend this insight to investigating beliefs of responders and community members, symbolic meanings in narrative accounts, and practices of disaster response organizations.

Finally, we should understand these levels as nested structures that provide an opportunity to trace a longer process of inequality by connecting shorter mechanisms occurring at each level. More specifically, a network perspective of social capital has tremendous leverage for uncovering informational and resource inequities that span urban spaces, organizations, and social groups (Breiger 1974; Granovetter 1985; Lin 2000; Lin, Cook, and Burt 2001; Small 2009a).

CHAPTER 2

Race-Class Logics of Urban Spaces

> There is a *deep* history of racism and violence and segregation in this area. And that's going to be here for a while. And that's not going to go away because a storm came ... When we're working to prevent displacement, we also care about these residents [*smiles, pauses*] who don't care about their neighbors.
>
> —Sapphire, thirty-eight, lead volunteer, Always With You

Through my observations and conversations with the founder and lead volunteers of Always With You, the local community-based organization in Eastville, I explore how the race and class boundaries marked in urban space seep into the disaster response process. From the onset, the placement decisions of the NGO Resiliency Is Us replicated the racialized spatial boundaries that divide Westville from Eastville. This in turn led to racialized zones of disaster response for Eastvillers and Westvillers, respectively. Resiliency Is Us situated one of its makeshift centers in an affluent residential cluster in Westville and set up its other location at the periphery of Eastville closer to Westville. Always With You, the local community-based organization, ran its disaster response center out of its storefront location in Eastville.

The sociospatial and sociohistorical cleavages of race and class of Westville and Eastville led to racialized zones of response. Race and class enter disaster response through a process that begins with the perceptions, beliefs, and practices of responders and survivors; racialized recruitment and voluntarism among volunteers; and most importantly through the placement decisions of large NGOs as well as the service perimeters of responders around the need for space and the concern for volunteer safety. The chapter focuses primarily on the latter.

Historicizing Race and Class on The Rockaways

The Rockaways separates Jamaica Bay from the Atlantic Ocean (figure 3). It has a combined population of 114,978 as of 2010. (See table 1 for a community profile of The Rockaways.) The western and eastern side of the Rockaways are highly segregated by race and class. The residents are stark opposites on measures of income, occupational status, transfer receipt, and owner occupancy. This segregation on the Rockaways leaves the residents residing in the eastern part of the peninsula socially vulnerable to disaster (see figure 4).

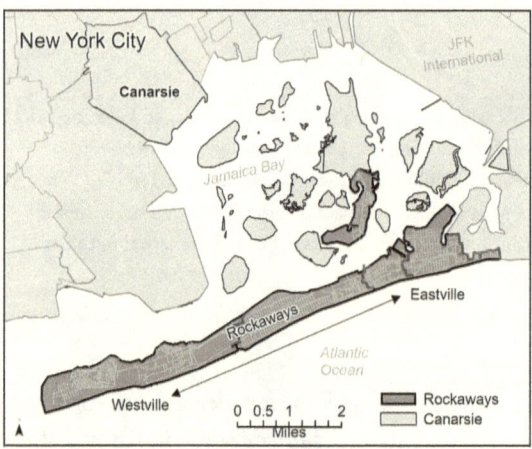

FIG. 3. Map of The Rockaways.

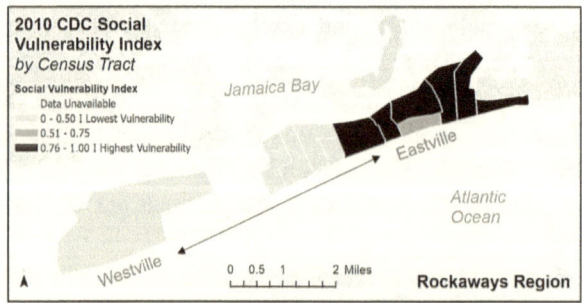

FIG. 4. Social vulnerability by census tract across The Rockaways. This map uses socioeconomic data, household composition and disability, minority status and language, and housing type and composition variables from the 2010 CDC Social Vulnerability Index to depict communities with high and low levels of vulnerability.

TABLE 1. Community profile of The Rockaways. The Rockaways consists of several zip code tabulation areas, so the ranges in column 2 represent the lowest to highest value across the ZTs in The Rockaways.

Community profile	The Rockaways
DEMOGRAPHICS	
Total population (2010 estimates)	114,987
Whites	49,088
African Americans or Black	47,957
Hispanics	24,102
Median age by ZT (zip code tabulation area) (2010–2014)	33–49
Percent of population that is male (2010–2014)	46.6–51.2
Percent of population that is foreign born or Caribbean born by ZT (2015–2019)	2.9–13.7
FAMILY STRUCTURE	
Percent of households that are female headed with children by ZT (2010)	2.8–24.8
POVERTY AND UNEMPLOYMENT	
Percent of families in deep poverty by ZT (2010–2014)	1.1–10.7
Percent of population in poverty by ZT (2010–2014)	3.1–25.2
Percent of population of Whites in poverty by ZT (2010–2014)	2.8–24.6
Percent of population of Blacks in poverty by ZT (2010–2014)	6.3–33
HOUSING TENURE AND TYPE	
Percent of household units that are renter occupied by ZT (2010–2014)	3.2–74.8
Percent of household units that are owner occupied by ZT (2010–2014)	25.2–96.8
Percent of household units that are single-family units by ZT (2010–2014)	22–94.5
Percent of household units with 2 housing units (duplex) (2010–2014)	1.8–24.7

After Sandy, dilapidated nursing homes, drug rehabilitation centers, and halfway houses were visible toward the eastern part of the Rockaways. The eastern part of the Rockaways warehouses the urban economically deprived and socially vulnerable: such as those living in single room and boarding house occupancies, and other shared living arrangements. There are also residential clusters of recently migrated, first-generation immigrants renting basements and store tops; and clusters of high-rise apartment buildings, public and subsidized, housing Black and economically deprived families and the elderly who live alone. Public housing on The Rockaways goes back to the Robert Moses' Urban Renewal projects. He bulldozed dilapidated bungalows and replaced them with what is today a concentration of high-rise public housing (Caro 2006). Contrasting to this, The Rockaways contain the only "neighborhood in the entire country that has a majority-Irish population" (Kliff 2013). These neighborhoods with a high concentration of Irish descent residents (Figure 5) range from being modestly economically privileged to being quite affluent.

FIG. 5. Ancestral demographics of Rockaway communities.

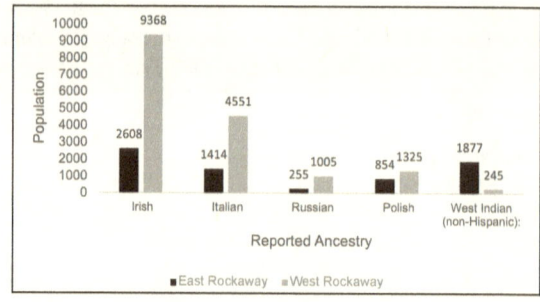

FIG. 6. Map of racial segregation on The Rockaways.

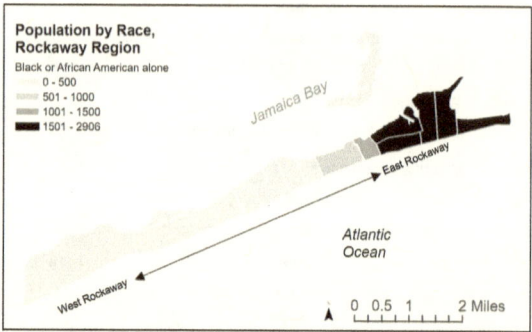

The race and class segregation of The Rockaways has a long history. From 1830, The Rockaways served as a seaside getaway for wealthy Manhattan businessmen and civil servants who came to escape the scorching heat of the summer (Bellot 1917). However, it eventually lost its appeal after World War II, leaving many of its summer bungalows vacant. The Rockaways's residents have always considered themselves separate from New York, and in 1917 they actually sought, albeit unsuccessfully, to secede from New York State on the grounds of disinvestment despite paying more than their fair share in taxes (Bellot 1917). New York City was a significant port, welcoming over one million Irish immigrants fleeing the potato famine in 1845 (Glazer and Moynihan 1970). While the Irish met with discrimination in cities such as Boston, in New York Irish immigrants soon became a politically formidable ethnic group (Ryan 1999). They also became dominant in civil service jobs such as those in the New York City police and fire departments. By the mid- to late 1800s the Irish were one-quarter of New York's residents.

The arrival of African Americans to the Rockaways was a highly racialized process. In 1950, the Welfare Department placed the most economically deprived African Americans who migrated from the South as part of the Great Migration, in The Rockaways. Commissioner Raymond M. Hilliard and the Welfare Department settled African Americans on public assistance, who, due to anti-Black racism, were "ineligible for public housing and not easily placed in private rentals." (Kaplan and Kaplan 2003, 69). The government incentivized small business owners of vacant summer homes toward the eastern part of The Rockaways to convert their homes to rooming houses to accommodate welfare clients (Kaplan and Kaplan 2003). In exchange they were to receive guaranteed government funding with little to no accountability, leading to providing inhabitable housing for economically deprived African Americans (Kaplan and Kaplan 2003, 69). This practice contributed to The Rockaways having the reputation of being a "repository for problems the city did not want to leave in the city's center" (69).

There is also a Jewish population among these nonWhite residents toward the East. The Black population had been placed in proximity to Jews because "the Irish population provided more resistance than Jews ... to the placement of Black tenants" (Kaplan and Kaplan 2003, 58). By the early 1970s, there was already an influx of Latinx and Caribbean immigrants in The Rockaways. Many of them are among the working and nonworking economically deprived residing in the eastern part of the Rockaways (see figure 6). However, a 2001 New York Times article referred to the westernmost part of The Rockaways as "the whitest neighborhood in the city, once known as the Irish Riviera [where] families go back three and four generations" (Scott 2001).

How Race and Class Organized Disaster Response

Sapphire and I stand beside a debris-filled, fire-obliterated lot in Eastville, The Rockaways. Sapphire is a thirty-eight-year-old Puerto Rican disaster response volunteer with Always With You. During Sandy's landfall, the flood waters triggered a fire that consumed three blocks of decrepit apartments stacked atop a row of immigrant-owned retail spaces and a storefront church where El Salvadoran immigrants worshiped. Sapphire describes one way that race and class intervene in disaster response; namely, through disaster survivors themselves:

> There is a *deep* history of racism and violence and segregation in this area. And that's going to be here for a while. And that's not going to go away because a storm came... When we're working to prevent displacement, we also care about these residents [*smiles, pauses*] who don't care about their neighbors. It's very, very tough to negotiate your place as someone from outside who is coming to do advocacy for residents. How do you even try to bridge communities that have been segregated from each other for this long? I went to a community meeting... and the way the people of [Westville] spoke about the people of this neighborhood [Eastville]. [*Sighs*] The contempt that they expressed about the people of this

neighborhood.... A fireman said the fire [which had consumed three blocks of primarily immigrant businesses, a storefront church, and apartments] did not go far enough.

Beyond the overt racism among disaster survivors and even responders, there were more subtle but important ways that race and class filtered into disaster response. Resiliency Is Us tended to attract White, non-Black, non-Latino, and middle-class nonlocal volunteers with leisure time or flexible employment and deployed them in neighborhoods with similar or different demographics. At the very least, disaster responders were certainly aware of the race and class similarity and differences among themselves and the communities in which they were stationed.

Alternatively, the racially minoritized tended to volunteer with local organizations in racially minoritized and economically deprived residential areas, particularly their own communities. For example, Sapphire, who was Puerto Rican and spoke fluent Spanish, continued to volunteer in Eastville because it had a "strong El Salvadoran contingency" that she felt especially obligated "to do advocacy for." She explains that the Spanish-speaking residents who spoke little English were "lost" after the storm. She further explains Eastvillers were "discarded" because of the neighborhood's extreme level of poverty and reputation for crime.

Race and class also organize disaster response through the placement of disaster response centers that adhere to race and class sociospatial boundaries both within and between neighborhoods (Charles 2003; Hunter 1974). For example, Resiliency Is Us, supported by several other NGOs from out of town, established its distribution center in the hub of Westville. Contrastingly, the disaster response center in the hub of Eastville was set up by Always With You, where Sapphire volunteered. There were no large NGOs in the visibly decayed center of Eastville, which Freddie, the founder of Always With You, described as where the government scattered decrepit nursing homes, boarding houses, and drug rehabilitation outpatient clinics and halfway or three-quarter houses. George, a fifty-eight-year-old unemployed Italian Eastville boarding house resident, describes other SRO neighbors as "mostly elderly, some are drug addicts, you know some work the streets, they're retired people, ex workers from the home—they ain't workin' no more." Although Resiliency Is Us set up its disaster response center in Eastville, that location was closest to the Westville border and inconveniently located away from where these most vulnerable Eastville residents lived.

The historical geographic distribution of wealth of the communities, which correlated with race and class of residents, also affected disaster center placement. For example, the Westville Resiliency Is Us center was conveniently located in the heart of the residential areas of Westville. Westvillers knew this location well and frequently traveled by it, as it was the best route to several stores. Contrastingly, the correlating street in Eastville used to house an immigrant storefront church, sandwiched within a row of retail spaces stacked with second-floor apartments, which burned to the ground during the hurricane. The location of the Westville center was in the neighborhood of third- and fourth-generation Irish residents, who are at the top of the racial-ethnic hierarchy both in terms of wealth and status. The community church was a Catholic church, which has historically owned large real estate and continues to own several buildings, but most importantly it owned vacant land, where Resiliency Is Us was able to operate. This is significant because large vacant land space is a prime asset during disaster response, since it is one of the primary factors in locational choices of large NGOs wanting to set up their response operation.

The way that race is spatialized in urban areas combines with racialized notions of "safe" and "unsafe" areas, influencing locational decisions of NGOs. Referring to the Resiliency Is Us's Eastville-Westville border location, Sapphire explained that the large NGO was told they should set up in places that are safe, and that the NGO "defined that as right next to the police precinct and that's where they operated." The Resiliency Is Us website confirms that safety is one of the criteria for selecting where to locate their disaster response centers.

Related to center placement, the narratives of volunteers that link poverty level and race-class composition of neighborhoods with threats to volunteer safety also influence the service perimeter of out-of-town NGOs. Sapphire complained about the Resiliency Is Us center in Westville and the racial dynamics that surrounded the placement of the center:

> When the Westville center was open, we heard from more than one volunteer that agencies would not provide services beyond a certain point on the peninsula—for safety concerns. Stories of a death of a volunteer—which has never been confirmed in any way but was only a folklore. We heard this from volunteers. Especially those coming from other parts of the country.

Sapphire's account illustrates the relevance of race, class, and poverty composition of urban areas to the placement of NGO centers in disaster areas. Her account also illustrates how perceived threats to volunteer safety may have influenced the service perimeter of the nonlocal NGO. Beyond talk, the impression of exclusionary placements and practices of the Resiliency Is Us disaster response center, held by the Always With You disaster responders, led to further entrenchment of segregating disaster response perimeters. Sapphire told me that this action by Resiliency Is Us triggered a counteraction by Always With You to establish their own response perimeter.

Sapphire stated that the Always With you decision prioritized Eastvillers in the way that she thought the NGO had been prioritizing Westvillers, to the exclusion of Eastvillers. This mutual enforcement of a disaster response perimeter leads to inequity in disaster assistance across these communities. This is because Resiliency Is Us has greater resources than Always With You. In turn this leads to greater race and class inequality across these racially and ethnically divergent communities where such a great wealth gap already exists.

In the following chapter, I present how FEMA, another arm of the disaster response machinery, similarly contributes to the reproduction of inequality based on ethnicity, citizenship, and legal status among Canarsie disaster survivors. Since Canarsie is homogeneously Black, it also provides a comparison of disaster survivors' experiences with FEMA in Westville and Eastville.

CHAPTER 3

Black Immigrants and Disaster Inequality

> I've been ordained forty-six years, and I have been here for the last twelve years. So, I've gotten to know the community, and it's a wonderful community to be in and it's a real mixed community. We have a Haitian community, we have all the Caribbean community, we have, um, people from the African, Nigerian, Ghanaian, Liberian, all over. So, it's a real mixed community. It's been a joy being here with them. But right now, there is uh—I would say three hundred or four hundred homes that have been severely damaged, I mean I think that's probably more than that. But I don't even know about it, you know.
>
> —Father Francis, Catholic priest, Holiness Church

While many immigrant adaptation studies focus on the segmented experiences of immigrants during routine periods, in areas such as employment and education, in this chapter I focus on the segmented experiences of immigrants during a period of urban disaster. I illustrate how the segmentation that occurs as part of the migration experience further intersects with experiencing a disaster. The chapter also extends the conversation about segmented incorporation into the spatial and housing structures of an urban community. This intersection of migration and disaster presents distinct types of challenges to various classes of Canarsie's Black immigrants.

Legal Segmentation of Immigrants

New York City is a well-known diasporic capital for Caribbean immigrants, where they are the largest immigrant group (Passel and Clark 1999; Foner 2005; Heron 2001). In 2005, 42 percent of U.S. Black immigrants from the Caribbean resided in New York (Kent 2007). Historically, New York has been the "principal gateway for immigrants entering the United States," making New York City one of the most diverse U.S. cities (Passel and Clark 1999). For this reason, classic works in urban sociology and immigrant adaptation such as Nathan Glazer and Patrick Moynihan's *Beyond the Melting Pot* (1970) have focused on New York as a site to study the adaptation experiences of new immigrants. Many Caribbean immigrants settle into various urban communities in New York City (Foner 2005; Heron 2001; Kasinitz 2008; Model 2008; Waters 1999).

Like many other immigrant groups, Caribbean immigrants undergo socioeconomic incorporation into the United States. However, many studies on Caribbean immigrants in the United States tend to exclusively focus on those immigrants who have maintained legal status, have moderate to high skills, and have gained employment in the formal sector of the economy (Kasinitz, Mollenkopf, and Waters 2004; Kasinitz 2008; Model 2008; Waters 1999). Many Caribbean immigrants migrate to the United States as nurses and teachers on H1B1 visas, making the socioeconomic adaptation of these groups of high-skilled immigrants a smoother process than for their counterparts who do not have these opportunities. However, most Caribbean migration occurs through family sponsorship. In these cases, migrating family members have a variety of skill and educational levels, but there is no guarantee of a job upon entry.

Like other immigrant groups, there are also Caribbean immigrants lacking legal status. The primary mode of entry into the United States for Caribbean immigrants is inspection via airports, where they are permitted entry upon presenting a visa (Foner 2000). This means that many Caribbean immigrants lacking legal status came in through an inspected port of entry but overstayed their allotted time in the United States. Some 33 to 50 percent of persons lacking legal status in the United States are persons who legally entered through inspected ports and overstayed (Pew Hispanic Center 2006; Passel and Cohn 2011; Warren 2003). However, White immigrants from Europe, Ireland, and Canada also contribute to the pool of persons who overstayed their visas in the United States. Approximately 30 percent of New Yorkers who lack legal status are from the Caribbean (Passel and Clark 1999).

Among Caribbean immigrants who lack legal status, those from Haiti are more

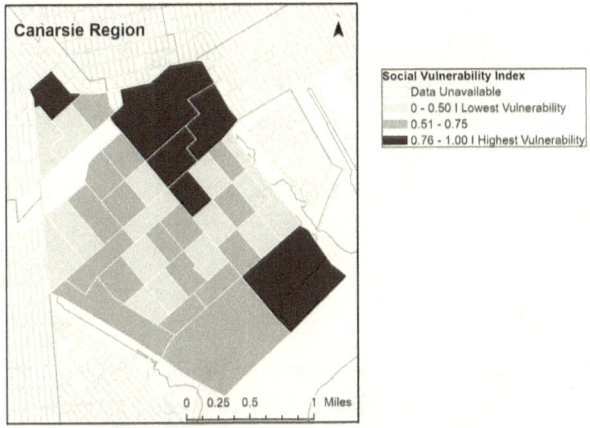

FIG. 7. Social vulnerability levels by census tract in Canarsie. Social vulnerability assesses a community's ability to prepare for and respond to hazardous events. This map uses socioeconomic data, household composition and disability, minority status and language, and housing type and composition variables from the 2010 CDC Social Vulnerability Index to depict communities with high and low levels of vulnerability.

likely to have entered the United States via boats through Miami, outside these inspected ports, precluding these immigrants from benefiting from even a temporary period of legal status. Furthermore, immigrants who enter the United States without inspection are excluded from provisions in immigration law that allow immigrants to "adjust" to legal status. By contrast, this opportunity is available to immigrants overstaying their visas. Immigrants lacking legal status participate in the informal economy, where women become babysitters, nannies, and housekeepers in White middle-class and affluent city and suburban households. Men go into trades such as carpentry, construction, and car repair.

This segmentation of the legal incorporation of immigrants into the United States (Portes and Rumbaut 2006) creates a corollary segmentation of the socioeconomic opportunities of immigrants in other realms. The precarious labor market position of many working-class immigrants coupled with limitations to governmental social support services makes them socially vulnerable to disasters, reflected in Canarsie's level of social vulnerability (figure 7). This reality is even more consequential among those immigrants who lack legal status and do not qualify for federal disaster support. Immigrant churches become a crucial source of social capital to help immigrants adapt to their unfamiliar environment.

Margarita A. Mooney (2009), in her book *Faith Makes Us Live*, examines the adaptation experiences of Catholic Haitian immigrants in the UK, Canada, and the United States. She finds that immigrant churches are committed to immigrants' adaptation and are particularly responsive to the hardships of immigrants lacking legal status. Churches also increase opportunities for networking, social support, and friendships for members (Krause 2008), all of which are crucial for improving their life chances.

Historicizing Race in Canarsie

Canarsie is a neighborhood in the southeastern portion of Brooklyn in New York City (see figure 8). Canarsie neighbors East New York, Brownsville, East Flatbush, Flatlands, Mill Basin, and Bergen Beach. Canarsie has a combined population of 97,137 residents as of 2010. Canarsie was not always the Black and immigrant enclave that it is today. The first West Indians to settle in Canarsie were seamen from Barbados, Jamaica, and the Bahamas who abandoned their ships in the early 1900s. These immigrants lived in squalid conditions alongside economically deprived Irish and Italians, as well as African Americans who had migrated from the South as part of the Great Migration (Brooklyn Public Library 2016).

These early Canarsie residents lived near Jamaica Bay and south of Colored Colony along Avenue J and K (Brooklyn Public Library 2016). The government razed their community and replaced it with a public housing complex under the 1950s "urban renewal" slum policy (Rieder 1985). However, this community did not become a Caribbean enclave until the 1965 Hart-Celler immigration law which precipitated a large influx of immigrants from the Caribbean and Africa.

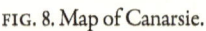

FIG. 8. Map of Canarsie.

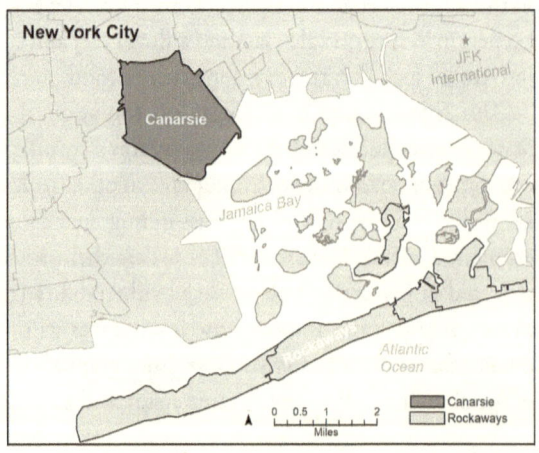

FIG. 9: Ancestral demographics of communities in Canarsie.

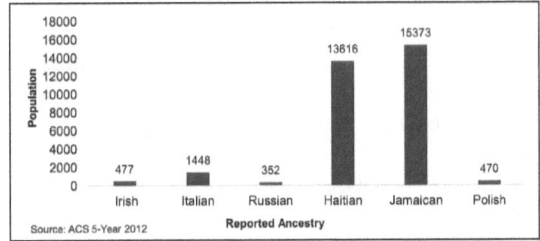

A snapshot of this neighborhood's immigrants in the 1960s would reveal Italians and Jews strolling the streets, instead of today's Jamaicans, Haitians, and other Caribbean and African immigrants (Rieder 1985). A key historical moment that marked the switch from yesterday's White ethnic enclave to this majority Black neighborhood is the violent opposition of Jews and Italians to the integration of Puerto Ricans and Black people into Canarsie (Rieder 1985). In 1972, when the school board ordered Canarsie to enroll a few dozen Black children into their schools, ten thousand White students boycotted school for a week (Rieder 1985).

Up to the 1980s and 1990s, Canarsie was still majority Italian and Jewish. However, from the 1990s, Jews and Italians eager to flee their changing neighborhood opened up the private housing markets for non-White residents (Rieder 1985). West Indian immigrants from the Caribbean, where homeownership is the hallmark of success, seized the opportunity for homeownership (Scott 2001). While many thought they were buying into integrated neighborhoods, they quickly realized that they were part of this ecological "succession" induced by White flight. By 2000, the Caribbean population had grown to six times its size ten years earlier (Scott 2001) (see figure 9).

Today, the majority Black Caribbean population in Canarsie, as in many other Brooklyn neighborhoods, enjoys tremendous political capital, having elected representatives at every level of government, including U.S. senator Yvette Clarke. The community has tremendous community social capital through its many civic associations (Scott 2001). (See table 2 for a community profile of Canarsie.)

Class and Space Segmentation

Canarsie has a Caribbean and Black diasporic community made up of multigenerational immigrant families. This means that while Canarsie is racially homogeneous, there is a lot of heterogeneity on the axis of class stemming from a refraction of migrant incorporation into the United States. During Sandy, many of these

TABLE 2. Community profile of Canarsie

Community profile	Canarsie
DEMOGRAPHICS	
Total population (2010 estimates)	97,137
Whites	7,219
African Americans or Black	82,588
Hispanics	7,884
Median age by zip code tabulation area (ZT) (2010–2014)	37
Percent of population that is male (2010–2014)	45
Percent of population that is foreign born or Caribbean born by ZT (2015–2019)	35.2
FAMILY STRUCTURE	
Percent of households that are female headed with children by ZT (2010)	19
POVERTY AND UNEMPLOYMENT	
Percent of families in deep poverty by ZT (2010–2014)	6.21
Percent of population in poverty by ZT (2010–2014)	14.5
Percent of population of Whites in poverty by ZT (2010–2014)	25.4
Percent of population of Blacks in poverty by ZT (2010–2014)	13
HOUSING TENURE AND TYPE	
Percent of household units renter occupied by ZT (2010–2014)	51.7
Percent of household units that are owner occupied by ZT (2010–2014)	48.3
Percent of household units that are single-family units by ZT (2010–2014)	22
Percent of household units with 2 housing units (duplex) (2010–2014)	50

immigrants of color had been concentrated in areas that were initially placed in Evacuation Zone B (see figure 10) but that became inundated with flood water, leading to being rezoned as Zone A. This means that they were at elevated risk of experiencing flooding.

The migration experience differentially affords opportunities and constraints for socioeconomic mobility. Migration experience and immigration status create a situation where immigrants, despite living in the same community, are differentially connected and integrated into the labor and housing markets of urban centers. As was evident in Canarsie, the differential incorporation into the socioeconomic space was reflected spatially in the urban environment as well. Different classes of immigrants were also reflected in the housing types and tenure as well as proximity to flood lines, which led to differentiated disaster experiences among this Black immigrant population.

Many of Canarsie's immigrant homeowner class with whom I spoke owned two-story, multi-family homes with basements and tended to be more established immigrants who migrated between twenty and thirty years prior to Sandy, when

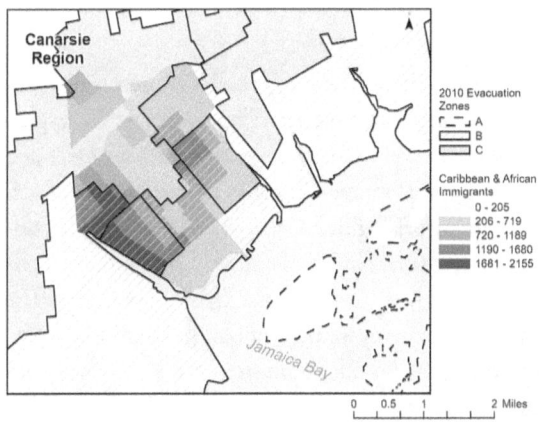

FIG. 10: Caribbean and African concentration in pre-Sandy Evacuation Zone B.

these homes were affordable for working-class immigrants. These immigrants had acquired legal permanent residency or naturalized citizenship status. More established immigrants were more likely to have had a chance to create large extended families in the United States. Many Caribbean families tended to live in the same household, community, or surrounding communities in Brooklyn. The longer their tenure in the United States, the more likely these immigrants had adult children who had legal status and had attained educational opportunities and more stable jobs in the formal U.S. labor market. This means that established immigrants with large families and legal status enjoyed the highest familial, economic, and social capital possible. Additionally, families who had lived the longest in Canarsie had the best opportunity to benefit from neighborhood capital in the form of involvement in party politics, neighborhood clubs, and churches, as well as interpersonal social capital through relationships with neighbors, coworkers, and members of their faith communities.

At the opposite end of the spectrum of physical, financial, political, and social capital were apartment renters and basement renters, who were usually those who migrated more recently and in some cases lacked legal status. Some immigrants and second-generation residents were young to middle-aged and had not attained desired educational and occupational goals. They lived alone or far away from extended family. Those with families were more likely to comprise younger heads of household with younger children who rely on, rather than provide for, the financial and social capital of their parents. Renters, differentially positioned than homeowners, had more precarious attachment to their urban communities. Basement renters had only loose connections to the community, had low-wage service jobs, and were self-employed, seasonally employed, or unemployed.

Immigrant Family Multigenerational Living

This Black immigrant community experienced unique challenges resulting from basement flooding. Many of these Caribbean Americans' homes housed intergenerational families, where the basements were very often living spaces for extended family members, friends, and nonfamilial renters. The lack of designation of Canarsie as a flood zone made it quite surprising that Canarsie residents would contend with up to nine feet of flooding of their basements. At least one building I observed even had a watermark up to the halfway point of a first-floor window. The multigenerational living arrangement of many immigrant families produced a particular nuance in the disaster experience for adult children of homeowners who occupied their parents' basement. For example, Carol, a middle-class Canarsie Sandy survivor, lived with her husband and four-year-old daughter in her mother's home. While standing on her driveway, I asked Carol about her losses:

> SM: What kinds of things did you lose?"
> CAROL: We had a very nice clothes closet downstairs made of nice wood, very expensive. Took our time and everything. That's gone. We saved the washer and dryer—that's a good thing. A couch, TV stand, TV, another TV. All the heavy things. The toy chest.
> SM: That was your living area?
> CAROL: It's like a rec room. My exercise bike. Mattress. It's like a guest area plus the refrigerator, stove, it's just—
> SM: Wait, you had your kitchen on this floor as well?
> CAROL: Yeah, this is my main kitchen but, in the summertime, when we were barbecuing with the sink and stuff you wash out and everything downstairs because I don't let people go on this floor.

Carol's family shared the first floor with her mom, but she and her nuclear family also used the basement for laundry and recreational purposes. These kinds of living arrangements resulted in significant losses for second-generation immigrant adult children who had acquired enough to furnish their living spaces but had not attained sufficient upward mobility to purchase their own homes. The flooding eroded whatever gains these adult children may have made toward their own socioeconomic mobility. However, typically these young families' displacement would only be in the form of having to move to a higher floor of the same building structure.

I asked Ferdinand, the FEMA site manager at the Canarsie disaster response center, how FEMA handled cases like Carol's where the basement held "extra" electronic devices, furniture, and appliances. Ferdinand emphatically stated:

They will not cover it, because it's extra. Because with FEMA, really what we do is try to put you in a safe, sound, and functional position—if you have like ten TVs, you only need one. But if you have a child that is going to school and they have a laptop, and it's for school, they can write a letter stating that they need it because, and it's for schoolwork. And even for a car, FEMA can assist you with one car. In some cases, if you have the need for two cars, you state it. You say, "Well, two parents. One has to go pick up kids or babysitters, the other one has to go the other way to work," so it's an exception. They will consider it accordingly.

This rationale does not recognize how the multigenerational immigrant family living arrangement complicates the notion of what is "extra." While furniture and appliances may be deemed unnecessary for the purpose of replacement, they actually belong to adult children who also do not fall into the category of renters and whose names may not appear on utility bills. However, these issues are miniscule compared to the hardships nonfamilial basement renters had to endure.

Homeowners and Landlords versus Basement Renters

Basement renters, who lost so much, experienced immense challenges in seeking FEMA disaster assistance. Ferdinand and I talked about the immense losses of the basement renters in Canarsie. I expressed my concerns about the displacement of disaster survivors who lived in basements as well as the loss of personal possessions of Canarsie residents who occupied basements:

> SM: Okay. So, I've been concerned about people who have been displaced in Canarsie, because I've gone to The Rockaways, and it's a little bit different in that not a lot of people occupied the basements. But here is very unique where you have a lot of people who occupy the basements, and because the flooding was to the basements, it's equivalent to losing a house, because it's your entire living space in the basement. So, I'm concerned about these people and how they're able to navigate. Is there assistance available for them?
>
> FERDINAND: Yes, there is assistance, because they can file a claim for their personal belongings and the furniture as long as they can prove that they live there by getting a statement from the landlord saying that they've been living there, or they can prove that they pay bills, or they have something proving that they live there. Sometimes they replace what they lost. They can submit their receipt and ask for consideration or reconsideration based on the situation, and the inspector will come over also. When the inspector comes, they look at the situation. They don't have to see a lot. They can just,

> by looking at it, they see by the depth of the water, they know how much damage that they can assess. So, from that, they can give them an average of—tell them how much they qualify for, because the inspector submits the result, and the reviewer looks at the figures.

The reality of the lived experiences of post-Sandy Canarsie survivors is that these Black immigrants incurred challenges uniquely shaped by their migration experiences. Their lack of familiarity with navigating the U.S. institutional apparatus of aid, combined with the "color blind," "culture blind," and "class blind" orientation of FEMA, among other bureaucratic inefficacies of governmental response, severely interrupted their receipt of disaster aid. Additionally, a more nuanced attention to the intersectional experiences of immigrants further reveals that basement renters in Canarsie experienced a particular unsung fate.

Among disaster survivors whose homes flooded, basement renters were the ones who were most likely to experience displacement suddenly and unexpectedly. This is because the flood waters engulfed their entire living space. I saw water marks that almost reached the ceiling in Canarsie. As a result, they were also the ones to suffer the highest proportional dispossession of their personal capital. Basement renters were most likely to experience disruptions to accessing social capital shared among neighbors, which was available only to those who had the option of remaining in their homes. They also lost dyadic or interpersonal social capital with friends. Yet they were also the ones who were least able to take advantage of much needed government-mediated disaster aid reaching their community. I spoke with Vidalia, a FEMA caseworker at the FEMA disaster response center in Canarsie:

> SM: Did you see differences in experiences based on the floors that people lived in?
> VIDALIA: Yes, definitely. Those that were in the basement, those were the ones that were affected.
> SM: And the first floor was not as much?
> VIDALIA: Very few on the first floor. Most of the ones that were flooded were the basements.
> SM: About how many feet of flooding did they tell you?
> VIDALIA: Well, we've had some up to six, seven feet of water.
> SM: Do you recall the different stories of the people who came in?
> VIDALIA: Some people were able to have the damage restored, repaired.
> SM: Who did these tend to be?
> VIDALIA: These would be the homeowners. And some people, because the

basement was so badly damaged, they were just going to put the property up for sale.

Unlike many basement renters, homeowners were more knowledgeable that FEMA would provide grant assistance, while renters were not as aware that they too qualified for assistance in replacing their personal property. Amelia, the HUD specialist volunteering at the FEMA-run center to provide housing assistance to displaced Sandy survivors, pointed out the inequalities in receipt of assistance between landlords who occupied the higher levels of buildings and the tenants who rented either the first floor or basement:

> AMELIA: There's some people that have landlords and everything was flooded out. A Haitian American tenant came in on November 26th, said that everything was flooded out. Then he was over here somehow because he called the insurance people. He found out that his landlord had been compensated, but they didn't take care of anything in the basement that was his. He was like, "Well, what about me?" You know the hurricane took place at the end of October. They even asked him for rent for the coming month. He felt like that was a slap in the face. He lost everything.
> SM: So, he didn't apply for replacement of personal items from FEMA?
> AMELIA: No. And he didn't have renter's insurance either. That was too bad for him.

Canarsie residents were differentially impacted depending on whether they were homeowners with access to higher floors or basement renters who had nowhere to go. What floor residents occupied and whether they were homeowners or renters were usually also a function of socioeconomic status, which was tied to whether they had attained legal immigrant status. The following exemplars explore the salient cleavages and nuances in how various classes of Canarsie disaster survivors experienced navigating FEMA's grant assistance program.

Slow Arrival to fema Disaster Response Center

In speaking with Amelia, an African American disaster responder who is working at the housing assistance table at the center, I became aware that from the vantage point of the responders, these Black immigrants were arriving at the center much later than expected.

> AMELIA: Well, I can give you my opinion on what I think is happening with a lot of the Haitians and Africans. This community from what I've observed is an

> immigrant community. I'm guessing that a lot of the people were not aware, they weren't aware of the services that were available to them, or maybe they didn't think that they were eligible to come here for assistance. And when they found out that they actually were, many of the assistance that we had for them, either the period had expired for them to come in, they were too late, or it was just too late.
>
> SM: What time schedules did they miss?
>
> AMELIA: I know that they were giving hotel rooms away for people that were in need, and a lot of those rooms were taken obviously. Hundreds of families that were affected by the storm and my guess is they didn't know that they could come here and speak to somebody.

Among the Canarsie disaster survivors who came in, there were those who had not registered for FEMA before coming. Amelia describes the complicated disaster experiences of a young Black immigrant family whose basement apartment was fully engulfed by Sandy:

> AMELIA: His wife had just had a new baby. He had a wife and two children under the age of two. They lost everything in their apartment. So, he wanted to know what we could do to help them. Not only did they lose everything, but the utility service company had also come into their apartment, shut down everything, because apparently salt water had gotten into the boiler system, because they stayed after the storm—the boiler system—then they asked them to move.
>
> SM: And they stayed?
>
> AMELIA: They stayed.
>
> SM: Did they tell you how many feet of flooding they had?
>
> AMELIA: Seven feet of water, so it went all the way up. And so, he's like, 'I have two young kids. They shut off the heat. They say it's not safe for us to live there. What can I do?' So, I set him up. Yeah, and his story was just so sad because he said I didn't know that I was eligible to get a hotel room. I spoke to the female representative and she's like, 'He wasn't even registered. He didn't even register, and now he's coming and asking for assistance. All the hotel rooms are already given away. In fact, some of the hotels are asking people to leave at this point, so he's out of luck.' And then I said, 'Well, okay, let me try and help you find an apartment." He said, 'I don't have money for even a down payment. I don't have anything. I was not expecting this. I wasn't prepared for anything like this.' I went through, and I was trying to find something for him that maybe he could afford. He just couldn't afford anything.

Time lags in obtaining information are costly and consequential during disasters, precisely because of these time horizons on programs. In some cases, there were repeated program startups, extensions, and deadlines. This situation created confusion and yet another set of delays, particularly if people were relying on word of mouth.

Even among those Canarsie disaster survivors who *did* register with FEMA and had received a voucher, there was no guarantee they would find an apartment to move into. These families' experiences exemplify the reality of many immigrant, economically deprived, and working-class families who are not financially prepared for disaster. Despite these immense hardships, once this resident was able to contact the official disaster response apparatus on the ground, all Amelia could offer was a referral.

No Rental Documentation

FEMA required basement renters to produce documentation proving that they lived in the basement. Many tenants did not hold leases either due to the landlord not having permission from the city to rent out the basement, which usually holds the utility room, or due to the tenant being an immigrant lacking legal status. In these cases, landlords typically do not take on the risk of providing proof that these tenants live in their basements. Trevor, the eighteen-year-old son of Canarsie homeowners, tells me about the gentleman who rented the basement from his parents and who became displaced and lost his job all at once:

> I don't know, but the thing was that he rented the basement without a lease, because it was their friend, so he couldn't get anything off of FEMA. He basically—I don't know if somebody put it in his head that it was a loss. I guess he didn't feel comfortable staying here anymore. My mom cut him that check, she wasn't like that, he was a good tenant, paid on time, so she gave him the money back, the security. And all that. And a little extra to get on his feet. That kind of thing. When the power went out, he was out of work. He seemed really stressed or anxious.

Amelia shared that most people who were seeking assistance were displaced basement renters.

> AMELIA: A lot of them that come to my table are from basement apartments. I would say the majority of them are basement apartments. I think all of them, actually, but I've talked to a couple of homeowners.
> SM: About how many cases do you think?

AMELIA: Well, I keep a log of that. Some days are slower than others. Twenty people if you count today. So, nineteen of those lived in basement apartments?

SM: And then, of those, how many would you say lived with friends versus how many were in hotels?

AMELIA: Not nineteen of them. Because a couple of them were homeowners, actually. Some of them were homeowners. Let me say, I'm guessing, maybe seventeen of them. And then everybody else is homeowners.

SM: Do you see any difference between what the homeowners and the people who live in basements experience? Anything that stands out to you?

[*Interruption*]

AMELIA: Some of the patrons are renters.

SM: You said that there were differences based on the floors that they lived in?

AMELIA: Yeah, because they were basement apartments.

SM: So, regarding the basement apartments, did you notice any differences among people who did not evacuate the basement before the storm?

AMELIA: A lot of them didn't evacuate because they didn't think they were going to be hit this hard.

SM: When the water receded, did they continue to stay in the basement?

AMELIA: No, they couldn't stay because of the mold problem. There were a lot of mold issues, so they couldn't stay.

SM: So, you didn't encounter anyone who stayed after the storm?

[*Nods*]

SM: Where are they now?

AMELIA: Well the homeowners stayed in their homes, but they just couldn't go down to their basements.

SM: What about the people who lived in the basements?

AMELIA: They rented the basements, so they don't have access to other floors. They had to find other housing.

Finding Housing

Amelia told me about her friend who had been displaced out of her basement apartment. I always welcomed these accounts because of the difficulty of running into persons who were displaced out of the neighborhood or who were still in shelters and had not come into the disaster response center for assistance.

AMELIA: I had a friend that was affected also, and she couldn't stay in the basement, so what her landlord did was she moved her to the top floor in one of her bedrooms, but she couldn't stay in the basement. They stayed until Rapid Repair came and fixed it up.

SM: Then she went back down?

AMELIA: She's not going back to the basement because there's still damage, so they took out all the mold and they took out the boiler and all that other stuff. The landlord is going to put the house up for sale.

SM: What's going to happen to your friend in the meantime?

AMELIA: She found another apartment to live in.

SM: Did she go to a basement? [*Chuckles*]

AMELIA: No more basement. [*Chuckles*]

SM: I'm going to ask you a little bit about your friend. What's her profession?

AMELIA: She is an insurance adjuster.

SM: So, she would know the process? Okay.

AMELIA: Yeah.

SM: How quickly did it take her to get another apartment?

AMELIA: I think it took her about a month, a month and a half. She found it last week.

SM: Did she remain in the same area?

AMELIA: No. She's in Flatbush. This happened over in Canarsie, Brooklyn, but she's over in Flatbush, Brooklyn, now.

SM: Did she ever tell you whether she wanted to be in Canarsie, whether she wanted to move away from Canarsie because of the flooding, or was it wherever she could find an apartment?

AMELIA: Well, I don't think she really wanted to move out of Canarsie. But because of the housing situation—because so many homes in that area were flooded—she just moved away from there so that if anything happened there, she wouldn't be affected.

In the above case, Amelia's friend benefits from both her human capital and her social capital through her connections. Amelia's friend was employed and had some knowledge of the housing market in the area. She also benefited from having a close enough relationship with the landlord to allow her to stay in a safer part of the building. However, in some scenarios people who occupied basements were also unemployed and did not enjoy this close relationship with their landlords or a local government worker and happened to volunteer at a disaster response center. Some were also unemployed before the storm, and others became unemployed as a result of the storm and faced double jeopardy.

Many basement renters encounter challenges with finding housing. Finding new housing is even more challenging for large families. Amelia, who also held a graduate degree in public administration, speaks about the housing discrimination that displaced disaster survivors were experiencing:

> AMELIA: Well, we work during disaster recovery, and as far as housing goes, we are trying to point people in the right direction. People that might have been affected by the storm that are maybe Section Eight voucher holders that are in public housing that might be homeless. Any sort of housing issue.
>
> SM: So, you were just telling me about the people who are homeless. Have you seen any?
>
> AMELIA: [*Nods*] There hasn't been equal opportunity as well, because we find that—we've heard that there are some stories of discrimination as well. Maybe a family tries to rent an apartment. They've been affected by the storm. They're looking for new housing, and maybe even an apartment. They're told 'I can't accommodate you guys because the family is too big.' Something like that.
>
> SM: Do you deal with cases like that? About how many cases you think you've seen?
>
> AMELIA: I haven't seen any personally, but I just heard that coworkers and other people working at the disaster response center. But that's something that many victims have been facing.
>
> SM: What happens to them, those victims who can't get permanent housing?
>
> AMELIA: They'll either end up in shelters if they can't find permanent housing. They'll end up in shelters. They'll go live with their family. They might just be homeless. But that's why we're here trying to address the needs and trying to help them find some sort of accommodation.

No Legal Status

Immigrants without legal status face an especially high burden when it comes to navigating bureaucracies even for lifesaving assistance. Therefore, during my conversation with Ferdinand, the FEMA site manager, I inquired about what was available to these immigrants through FEMA. Although he never said this, the conversation exuded a "don't ask, don't tell" approach and the usual refrain of no one is calling Immigration and Customs Enforcement that one hears whenever they press organizations about the welfare of those who lack legal status. Ferdinand also surmised that immigrants lacking legal status are not necessarily left behind because they also belong to mixed-status families, where the head of household qualifies. In this sense they indirectly benefit from assistance. Beyond the question of ineligibility, the fear of seeking assistance is a huge challenge among this population of disaster survivors and those hoping to serve them. Our conversation challenged both of us to consider the invisible barriers to seeking assistance that bureaucrats are not always perceptive to. I engaged Ferdinand, a Caribbean American, like myself, and someone who has deep insight into this community, in deeper discussion:

> SM: So, for people who lack legal status, and they don't have any U.S.-born family member in the household that qualifies them for FEMA assistance, what happens to them?
>
> FERDINAND: They can register. When they call the helpline, they will ask for a name. They're not going to ask for status.
>
> SM: Social Security number?
>
> FERDINAND: They will ask for—yeah, they need that.
>
> SM: So, that's a problem.
>
> FERDINAND: Definitely.
>
> SM: Because if you don't have status, then you don't have a Social Security number.
>
> FERDINAND: Yeah, we can't do anything.
>
> SM: That's the grey area. Do you see a lot of that happening? Where people don't?
>
> FERDINAND: No, not since I've been here, no.
>
> SM: So, I think some of what's happening is that there's already filtering that occurs even before survivors get here.
>
> FERDINAND: Yeah, probably.
>
> SM: People who don't have it don't even come.
>
> FERDINAND: Yeah, well let's put it this way. If they're here, they are functioning. There must be somebody that they can go under. I mean, they are not out in the cold, because I let them know and I tell them. Tell them, don't be afraid. We have nothing to do with immigration. So, if they feel comfortable, let them come here, and I'll talk to them. Don't be afraid. Don't be afraid because we're here for you. We're here for you. Don't let people scare you and tell you, 'Yeah, immigration.' We have nothing to do with immigration.
>
> SM: Right. I think they won't come, especially because it says 'Federal Agency' [*as I point to a logo*]. FEMA is also a 'Federal Agency.' 'Homeland Security.' [*I read aloud*]
>
> FERDINAND: It says that? Where?
>
> SM: Yeah, in this line.
>
> FERDINAND: Oh, no.

SM: It's right there.

FERDINAND: That is something. That is true. That is true. I didn't even think about that. Oh my gosh. But hopefully some of them don't understand what that means.

SM: Well, at least, it's really important that they've tried to put people here who have the same cultural background or nationality.

FERDINAND: Yeah, but it wasn't planned that way. They didn't know that I had the same background, because they were going to open a center. I don't know where they were going to send me, but my plan wasn't to be here. Not here. So, when I saw it, I said, 'Oh, that's my [old] neighborhood.'

A number of Canarsie Sandy survivors learned about the disaster response center through word of mouth. The fact that Ferdinand had a shared social identity, was from the community, and openly welcomed disaster survivors may have attracted some to the center who may have otherwise stayed away. However, the elephant in the room is that both FEMA and Immigration and Customs Enforcement (ICE), which executes deportations, are under the authority of the Department of Homeland Security. Therefore, it is not unimaginable that mixed-status families could see this connection as a legitimate basis for not seeking critical disaster recovery resources.

The Post-Katrina Emergency Management Reform Act of 2006, despite several amendments to the Homeland Security Act of 2002, kept the Federal Emergency Management Agency (FEMA) within the Department of Homeland Security (DHS). This potentially causes a hindrance for mixed-status immigrant families who are experiencing hardship but fear ICE. Although immigration enforcement was not a priority at the disaster response centers, the symbolism of the DHS signification can lead to unnecessary ambivalence to seeking needed disaster assistance for qualifying members of mixed-status families.

CHAPTER 4

Labyrinth Bureaucracy

> Well, we're the forgotten residents of Brooklyn. Still waiting to get some help. You know, I'm still going back and forth trying to get all the paperwork done so I can reapply for FEMA and try to get assistance. You know, because the money I've received from FEMA is not enough to take care of the damage, not even half of the damage.
>
> —Sherri, thirty-eight, high school teacher, Canarsie disaster survivor

The Canarsie disaster survivors I met and spoke with felt inaudible and invisible, so they were eager to share their experiences with me in the hopes that I would relay their hardships in this book. These disaster survivors had not been prepared for flooding and did not receive timely evacuation warnings. Canarsie had been zoned as Zone B (see figure 11), and New York City would only designate Canarsie as Zone A after Sandy (see figure 12). I relay some of the conversations I had with disaster survivors while standing on the street corner, climbing up their front steps, and sitting at their kitchen tables. Some conversations occurred while walking through the basements of those survivors who agreed to talk with me while I canvased the areas that I heard had experienced the highest levels of flooding.

Many of the Canarsie Sandy survivors conveyed that they wanted others to know about their experiences in the hopes that these revelations would lead to future changes in the execution of disaster response. These are the unheard voices and lived experiences of Canarsie's Superstorm Sandy survivors. These exemplar dialogues present salient cleavages in the disaster experiences in this New York

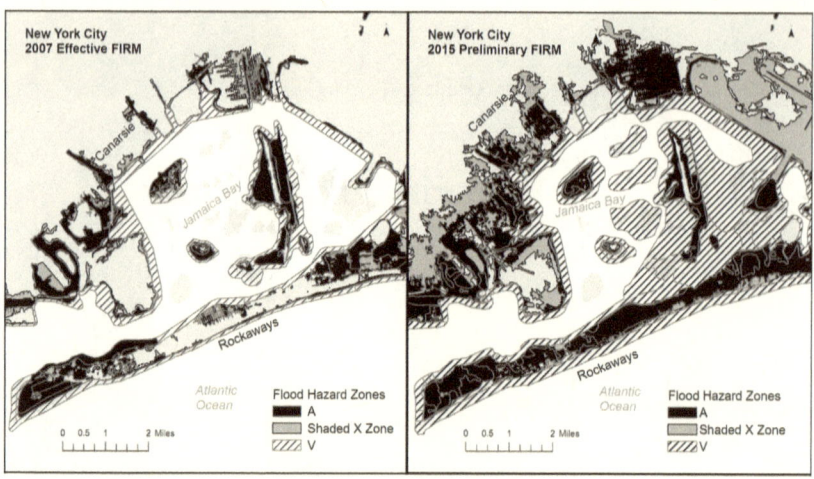

FIG. 11. Flood zone designations for Canarsie in 2007 and 2015.

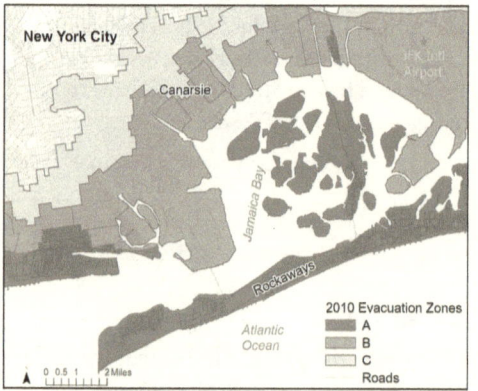

FIG. 12. Map of Canarsie pre-Sandy Zone B designation.

City Black immigrant enclave. These conversations with disaster survivors served as a form of catharsis for them, even while they were still in the throes of crisis. Together these dialogues help elucidate the various mechanisms and points of neglect, confusion, frustration, and inequity they felt and experienced during the process of awaiting, seeking, delivering, and receiving state and federal governmental FEMA disaster assistance.

The Forgotten Little Town

Twelve days after Sandy's landfall, we arrived in Canarsie, where Sandy had flooded several basements and some first-floor apartments and homes. Within a few minutes of our arrival about sixty to eighty middle-aged to elderly Afro-Caribbean survivors, about three-quarters of them women, some with children, surrounded the open back of the truck. Almost two weeks had gone by since Sandy struck, and these Canarsie disaster survivors told us that they had never received this type of assistance. They partook in the food, baby formula, diapers, flashlights, batteries, blankets, winter jackets and coats, clothing for infants, children, and adults, and so much more that my family and I collected through our disaster response drive in North Carolina.

Debra, a fifty-five-year-old single working mother of two school-aged children, who had been displaced from her home, agreed to talk with me. She stepped away to retrieve donated items from the truck, but she was still within an earshot of her niece and me.. While I awaited her return, her niece, Alma, who accompanied her, could hardly contain her exasperation.

> ALMA: Okay, I'm with my aunt. She lost everything in Canarsie. Her whole basement was flooded, and my cousins lost everything that they have. They don't have light, no heat, and she's, like, sleeping from house to house. They can't—and she's a working woman, and no help. Like, they still don't have lights. And what today is? Like the eleventh day after the storm?
>
> SM: And has she been trying to call?
>
> ALMA: Yes, and she's been getting like, you know, those automated message systems. Yeah. They tell her that the lights were coming on last night. It didn't come on.

I began to ask Alma whether they had heard of any other communities that had received the types of assistance they were still lacking, but Debra, who had been listening from afar, quickly interjects, "I—I think in Red Hook, they were giving something. But in Canarsie, apparently, we are the forgotten little town. Nothing, nothing, nothing. I am still without light."

Red Hook is a majority White urban area in South Brooklyn and Canarsie a majority Black immigrant enclave in Southeast Brooklyn. Debra further explains how the disaster led to displacement and splitting up her household as she navigates seeking health and educational resources for her daughter.

> DEBRA: I have two asthmatic children who need to be—you know, get medication and stuff. I had to have one girl stay with someone else. Every day she's crying. She doesn't want to go to school because she's so depressed. We saw the water. She was trying to help us with it. It is really, really unbelievable—tumultuous.
>
> SM: And then you've been trying to call FEMA?

DEBRA: I've been calling and calling every day, and calling just now, they start hanging up on me.

SM: Oh wow! What would you attribute to this lack of response, or lack of help?

DEBRA: Apparently because we weren't zoned as zone A, they cannot respond to us in time. I guess if we were in Zone A, then they may respond more urgently. Last night the NYPD [New York Police Department] came, and they put some light, because we still are out of lights on this avenue.

SM: "Mm-hmm."

DEBRA: I don't know. That's the only thing that they did, so far.

SM: Okay, but it's still cold?

DEBRA: My boiler's gone. My, um, my water heater's gone, so we have no hot water, no heat, nothing.

SM: And the children, how are your children?"

DEBRA: Well, I've—the school called, and the school is going to provide some comfort to one of my kids here, because she's very stressed. She doesn't want to hear the water. She doesn't want to see the water. She's traumatized.

SM: Sorry to hear that.

DEBRA: Yeah. That's where we are right now.

SM: So, you think the town was forgotten because of how it was zoned?

DEBRA: It was zoned, exactly. Uh, we're sitting in a basin because we have three bodies of water around us. So, how could it not have been zoned? Oh, yeah, we need blankets. [*She runs off again*]

As with Debra's experiences of trying to mother while displaced, Megan Reid (2012), in her disaster study on Black economically deprived and single mothers, found that these already marginalized women experienced additional disaster-related hardships because displacement impacted not only them but also their ability to effectively mother their children. It is particularly difficult catering to the psychological needs of children traumatized by disaster.

In the face of institutional failures, racialized single parents, usually mothers who are burdened with expectations of child rearing, must rely on kin and friends to temporarily buffer the social support that is immediately lacking from the State. However, Alma's statement at the beginning of our conversation that Debra was moving "from house to house" reflects the tenuous quality of support from social ties of the economically deprived and working classes (Smith 2005; Desmond 2012). This pattern of fleeting bursts of support from informal networks of the economically deprived and working classes is a recurrent theme both in Canarsie and in low socioeconomic status (SES) Eastville Rockaway. In other words, although the social networks of the economically deprived are available, their social capital, in the form of a steady supply of resources in this case, gets constrained.

FEMA, New York State, and New York City had collaborated to quickly deploy teams that walked the neighborhood blocks, spoke with disaster survivors, and cleared fallen trees and debris. However, although many of these disaster survivors had learned of FEMA assistance and had registered with FEMA early, many anxious disaster survivors experienced interruption in receiving assistance because of a breakdown in the follow-up mechanisms of FEMA's application process.

No Response

A small-framed, soft-spoken Black woman, wearing a lilac floral-patterned dress and a snugly placed hijab framing her face, agreed to talk with me while she waits in line to receive items from the back of the truck. Natasha is sixty years old, a single working mother of five children plus one grandchild, all of whom live with her in an apartment down the block from where we were standing. Natasha shares with me her dire circumstances with food insecurity and discomfort, living with young children, in a home that is cold, damp, moldy, and filled with debris due to the flood. She explains that despite her efforts to reach FEMA while still having to go out to work, she has not received a clear response from FEMA as to what assistance she could receive.

> NATASHA: We're totally flooded out. We have no hot water or heat and I've been trying to get help from, like, FEMA, but they don't respond back. I don't know, some people might have interviews, some people don't have interviews, and I'm still waiting.
> SM: Okay.
> NATASHA: Like, my backyard is filled up with a whole lot of garbage.
> SM: Alright. [*I nod and maintain eye contact, expressing my deep empathy*]
> NATASHA: I'm not sure if we can use that dump today. [*She points to a dumpster*]
> SM: Alright.
> NATASHA: It's like hard getting out. I have six children in the house. And we're just like cold.
> SM: And how is it for them?
> NATASHA: They are trying to help out, you know. It's just a lot of crowd and mess everywhere. It's very uncomfortable, and I still have to go to work.
> SM: So, when you've sent them [FEMA] an email, they don't respond to you?
> NATASHA: They respond, but they don't tell you—they say wait for an appointment for an inspector to come out. But if they don't have an inspector

coming out within seven days, make contact with them. [*Sighs, pauses*] Call the number back—oh, we're busy.

SM: And what do you attribute to the lack of response to your neighborhood?

NATASHA: I think they forgot about us, but I see someone didn't. [*She smiles, referring to our presence in this impacted area*]

SM: Well, we came out here because we were looking for neighborhoods that were thought to be neglected. We spoke with Bishop Fabian, and then he was like, 'Yes, we're out here. Come over here.'

NATASHA: That was nice, thank you. I'm just thinking that somebody—it's over a week later. You know what I mean? Going through a whole lot more, like food spoils, and then when they said something about food stamps, they did not respond to that. They say they want to help you.

SM: What did they say?

NATASHA: I saw an email about anyone who's in the area that was affected by the storm would get assistance, but I think that's still on food stamps. Our whole freezer went out, and we lost all our food. We're still waiting. And they still didn't get us anything—so I don't understand what they're helping with.

SM: And you've been calling and nothing. And you have, you said, six children.

NATASHA: Right. An infant, my granddaughter. That's six children plus myself.

These economically deprived, working women's social vulnerability to disaster related to the challenges they experienced in carrying out their responsibilities as mothers while also trying to secure disaster resources (David and Enarson 2012). Natasha's experience with food insecurity and her dilemma of who is responsible for replenishing spoiled food when Supplemental Nutrition Assistance Program (SNAP) benefits have run out foreshadows the next chapter's question: What happens to the social safety net for the traditionally economically deprived during disaster? Similar to Natasha, Megan Reid (2012) found that economically deprived mothers had difficulty navigating the service bureaucracy of securing access to food stamp benefits. These women also had difficulty enrolling children in schools in their relocated neighborhoods. Economically deprived working mothers are breadwinners who are unable to take time off from work due to being primary caregivers to their families.

Economically deprived and working-class Black women are constantly and actively patching together support during disaster, as their networks transfer information, tangible assistance, and emotional support (Reid 2012). These kin networks help them and their children with evacuating to safety as well as processing trauma (Reid 2012). Single mothers rely on informal networks to alleviate transportation and childcare needs (Reid 2012). However, economically deprived women's networks, because of insufficient resources and limited geographic span, can leave women's recovery at the mercy of assistance from strangers and from disaster response organizations such as FEMA, large NGOs, and churches (Fussell 2012).

Black middle-class Caribbean American women similarly experienced communications issues with FEMA. However, they did not experience the kinds of hardships that economically deprived single women faced. Carol, a middle-class, second-generation married mother of one child who shared a home with her mother, who owns the home, stood with me outside on the driveway of a two-story, single-family home. It had been a couple of weeks since Sandy, and Carol was frustrated that she had not been able to fill out an application with FEMA due to loss of internet connectivity. She explained that she was only receiving automated messages that didn't allow her to reach a live person or leave a message. She demonstrated what she calls the "vicious cycle of hearing everything," the nonresponse that led to her immense frustrations with FEMA. Carol quickly whipped out her cell phone and pulled onto the screen a saved contact, an action she had performed several times before. Like a teacher, standing before a class, she held out her phone, "So this is the number that they give you." We both listened to the prompt from the FEMA hotline:

> You have reached the Federal Emergency Management Agency. FEMA's individual [*inaudible*] program does not provide food assistance or a $300 food voucher. Other organizations around your area may be able to help you with your food needs. If you have a food-related need please hang up and dial 2-1-1 for referrals to organizations that can assist. If you have losses other than food, including damage to your home, personal property, or vehicle, we recommend that you complete an application. Please stay on the line if you would like to complete a FEMA application assistance— [*Message abruptly cuts off*]

Carol expressed her frustration:

> CAROL: So, there you have it, they hung up. So, this is all I have been getting. And actually, when I went on Sunday, they told me I can register with my smartphone, because I didn't have the internet. And they have the call centers there, but they won't use the phone! So, I called, and it's been like that, like the vicious cycle of hearing everything. It says to stay on the phone—so there's nothing to press—you can't press anything. She still talks, and that's it.
> SM: So, it doesn't work?
> CAROL: No.

SM: And the only way you could have accessed it is by having a smart phone where, if you are able, to fill out an application online?
CAROL: Exactly.

Tracing the reproduction of race and class inequality should consider the ways in which Black and economically deprived women experience challenges stemming from how they are situated to both the pre-disaster and disaster response economy, including housing, health care, employment, and related benefits (Luft 2016). The experiences of both Debra and Natasha illustrate how disaster compounds the women's breadwinning, caregiving, and traditional domestic work responsibilities by adding the need to clean debris, navigate the loss of utilities and large home equipment, and seek assistance from their networks of family, neighbors, and friends, while trying to navigate the bureaucracies of service and disaster response organizations.

FEMA-to-SBA Labyrinth

My conversations with Ferdinand, the FEMA site manager at the Canarsie disaster response center, corroborated what I had been learning in Canarsie and The Rockaways. Our conversations provided insight into the on-the-ground implementation of FEMA's grant application and appeals process from a different vantage point. While FEMA made some strides after Katrina, Black residents I spoke with were caught in what I describe as a bureaucratic labyrinth in the execution of FEMA's disaster response programs.

After Superstorm Sandy, FEMA spent $25.5 billion in recovery funds in New York and New Jersey (FEMA 2017). By August 2014, FEMA had provided 183,000 applicants with $1.4 billion via the Individuals and Households Program (IHP), which I and Canarsie disaster survivors call the "FEMA grant." These FEMA grants assist with home repairs and rental assistance (GAO 2015, 9). Canarsie disaster survivors whose FEMA applications were denied are routed to the Small Business Administration to apply for a loan. This step caused much confusion and frustration especially for retired and fixed-income homeowners. These disaster survivors saw SBA as an unwelcome barrier to obtaining a FEMA grant. This sentiment is in stark contrast to how FEMA views SBA. A December 15, 2018, press release by FEMA describes SBA as "the federal government's *primary* source of money for the long-term rebuilding of disaster-damaged private property."

Although Congress passed the Post-Katrina Emergency Management Reform Act of 2006 to avoid the delays and coordination pitfalls during Katrina, themes of neglect and inequality permeated the narratives of the Black immigrant population of Canarsie. The Canarsie Sandy survivors with whom I spoke were already navigating the shock, destruction, disruption, and dispossession that the disaster event produced. Then they had to simultaneously navigate the unclear path to securing adequate and timely governmental assistance. Upon coming to the disaster response center, Canarsie homeowners, many who were either retired or minimally employed, were often diverted to apply for an SBA loan. Some of these homeowners were of working-class backgrounds, retired, or near the end of their mortgage commitments, and they could not fathom taking on a loan that they would have to pay back. Many of these Sandy survivors reported feelings of abandonment, frustration, confusion, exhaustion, and even distress, which were quite palpable as I patiently and empathetically listened to their stories. These residents attributed their distress to having to deal with FEMA. Some residents even spoke about abandoning their idea of seeking government assistance altogether, deciding it was best to attend to their needs on their own or rely on a patchwork of private contractors, handymen, churches, and family.

In my conversation with FEMA site manager Ferdinand, I gained deeper insight into the FEMA grant appeals process, which opens up the possibility of either increasing a disaster survivor's grant amount or changing a denial to an acceptance. One caveat was that the survivor who receives a denial or insufficient funding gets bounced over to their insurance company and the SBA before they can proceed with the FEMA appeal. The problem is that this method benefits those who are most adept at navigating the maze of such bureaucratic hurdles and those who are in the position to take a loan if it came to that. However, the Canarsie residents with whom I spoke at the disaster response center and out in the community were already quite frustrated with the delays they had experienced with FEMA. Therefore, receiving a denial or an amount that was significantly lower than their estimates for repairs and purchases caused them to give up prematurely. FEMA's procedural process of elimination looked to many like an opportunity for the government to derail them from a path of getting a grant, which they would not have to pay back, in an effort to lock them into an SBA loan that they could not afford. Ferdinand explained the FEMA-to-SBA connection to me:

> FERDINAND: Well, SBA is not really—don't go by the name, because it says 'small business,' it's a disaster-related loan. I thought the same thing when they said, 'Well, we'll refer you to SBA.' I said, 'What do you mean SBA?' I don't have a business, and I said, 'I'm not self-employed,' and they said, 'No, it's a low-interest loan, disaster-related.' I said, 'Oh, okay.'
>
> SM: An attorney in The Rockaways, told me that a lot of people on the more economically deprived end of The Rockaways were denied assistance. Do you keep a tally of the percentage of denials per community?

FERDINAND: We don't have the number, but sometime it will come out. They will put it out in the media, because, right now, they're still gathering information. So yeah, it will come out sometime.

SM: It will? Okay.

FERDINAND: It will come out, but what happened is there was a misconception when they said, 'Well, I've been denied assistance.' Most of the time that's what confused them.

[Ferdinand, continues to explain quite expressively]

FERDINAND: No, no, no, no, no. It's based on the primary contact. They will say ineligible because of insufficient damages. I don't like it. It doesn't sound right, but that's the way to say that something is not right. So, what they do is, they send you a letter, 'Right To Appeal.' When you appeal, if the inspector said there is insufficient damage, there can be two or three things causing this. It can be insurance. You said you have insurance? So, we stop right there waiting to clarify that. If you have insurance, your insurance was supposed to cover it. So, now you need to prove that you had no insurance. It's either you get a letter from your insurance saying, 'Sorry, you didn't have flood coverage,' or if there is a settlement, it says, 'Well, you are under-covered. You didn't have enough coverage. This is what we give you.' Now, you write back to FEMA. You appeal to say, 'Yeah, my insurance gave me that, but this is how much damage I have.' So, what they'll do, we will ask you to provide an estimate from the contractor.

SM: Okay. So, you have to get a contractor.

FERDINAND: Yep. So, once you have that, then look at what the inspector says, and what you said, and what the insurance gave you. They review it with all [this] new information, and they will come up with that, okay?

SM: But some people will appeal, and some people won't, and what I've been seeing is people who have more education, for one—higher income or whatever it is—they'll yeah, they'll go out and actually go and seek that information. Some people, when they get a denial, they don't want to go through all that. That's a psychological stress.

FERDINAND: Right, right. So that's why when they come here, that's why we said, 'It's not the end. Appeal. Appeal. Appeal.' I think you can appeal up to three times.

SM: Three times. But there are a few people who appeal, right?

FERDINAND: Oh, yeah. They do appeal. They appeal. That's what keeps us rolling right now, because we have a lot of people coming back who are unhappy about the result of the SBA. Because if you don't do the SBA, sometimes you get a little bit, but you could get more, if you do the SBA. Because if you do

> an SBA, and are not qualified, they decline your loan. That's a good thing in a way. That kicks you right back to the grant program. So, they will consider it. It is okay. Forget the loan. You can't afford it. You have no insurance. Now, I'm going to take care of you.
>
> SM: This favorable outcome depends on who comes in and whether they follow through with the appeal. Additionally, those who are qualified for the SBA loan will not be kicked back to the FEMA grant.

What Ferdinand describes is legible to those who have full understanding of how each part of the process connects to the other and are certain what the end will be. In reality, it is quite an arduous, time-consuming, uncertain process for people who are already under immense stress from having their lives so fundamentally disrupted. Additionally, during a disaster, time is of the essence, because people's circumstances are constantly in flux. For example, untimely assistance resulted in much more complications such as the growth of mold spores in basements of disaster survivors as well as the accompanying stress of dealing with it while thinking about the health complications one might suffer.

Another compounding issue is that after a long time lapse, it is possible that disaster survivors realize they don't qualify for any assistance from FEMA, SBA, or insurance providers. At the same time, the community assistance from local churches or community-based organizations may be stretched thin so survivors may be pushed further to the back of the line, despite beginning the process of seeking assistance early. Marlene's case exemplifies this. I met Marlene, a seventy-three-year-old retired seamstress from Jamaica at the disaster response center. Marlene is a Canarsie homeowner. I approached her for an interview in early December, and I later worked to assist her in getting connected with some services. In a follow-up visit to her home in March 2013, I wanted to see and hear how she had been coping since we last spoke. We talked as we walked through her basement, where only some Sheetrock hung on the walls. I asked Marlene,

> SM: Can you tell me a little bit more about your experience with FEMA? Because you said that they referred you to SBA.
>
> MARLENE: FEMA referred me to SBA, and a gentleman came by. He told me he's from the SBA. He didn't tell me he was from FEMA. Now, I'm naive to all this. I think the SBA was something different from FEMA. I didn't know that they were connected. So, when he told me he was from the SBA—he came, he measured, he did not ask any questions. He didn't ask, 'Did you lose anything?' Or 'What did you lose?' He just measured, cracked a few jokes, and left. He did give me his business card and his number—I think that's some number for his business. After that, I waited a long time, not hearing

from SBA, so I called and was told that there are so many applications that they're running behind.

SM: What dates was that—around what date?

MARLENE: I can't remember.

SM: That's okay.

MARLENE: I have everything written down. But they said that they passed some of the applications—including mine—over to another gentleman who would review it and give me a call. But before all of this, I did get a call from—I think it was Mr. Howard. He wanted to know why he was just hearing about my situation, because since I had called them somewhere around the 3rd of November, he wanted to know why my files were just reaching his desk when it was, like, late. Anyway, we talked, and he told me that somebody would get back to me. The person who got back to me was from the Small Business Loan about a month ago. This gentleman called me. He told me that 'Your loan application looks very good. Your credit is good. The only thing I see on it is a student loan.' They say you're not to give information when you're not asked. And I—my foolish self—said 'A student loan—it's not really my loan, it's my granddaughter's loan.' The gentleman, after his conversation with me, hung up the phone and called me back immediately, and he said, 'Ms. Marlene, I'm sorry the loan is denied because of the student loan. It's like ninety days late and it's a federal loan. I'm sorry I can't help you.' Then another gentleman called, and he now told me that he wanted me—he wanted to come by and inspect the place because they'll give me a grant.

SM: That's from FEMA?

MARLENE: From FEMA.

Consistent with Mr. Ferdinand's explanation, this funneling to SBA after a rejection of one's FEMA application was just procedure. Those who went through the process, only to get a denial from SBA, could then return to FEMA with their denial confirmation and use this as a basis for their appeal. Marlene's FEMA application eventually received a denial for other reasons, but only after she had expended considerable time, emotional energy, and effort. She had hoped that if she had gone through the process of going to FEMA, her insurance company, SBA, and finally back to FEMA, there would be a positive result.

The financial reality of many homeowners I met and spoke to in Canarsie reflected that the income of more than one working adult secured the mortgage payments. Homeowners either worked multiple jobs or had adult children living on the same floor with parents or renting out rooms or their basements. Those who had the most security before Sandy were those who had owned their homes for almost thirty years. However, these most settled immigrants, who tended to be first-generation immigrants approaching retirement, also feared that Sandy could plummet their equity or force them into losing their homes to mold if they were not able to obtain a FEMA grant.

Some disaster survivors I spoke to who had received a denial and diverted to SBA were adamant that they would not apply for a loan, completely unaware of this FEMA-to-SBA-to-FEMA labyrinth. This meant that this practice only served as a deterrent to completing a successful FEMA application for these Black homeowners. We see this in the case of Joseph, a sixty-four-year-old, Canarsie disaster survivor who migrated from Haiti and was approaching retirement. When he did receive a FEMA grant, which fell below his contracting cost plus his cost of replacing his car, he was rerouted to SBA. His SBA application was successful as he was offered a $46,000.00 loan. Joseph refused. However, Joseph did not accept this offer. Joseph and I discuss his experience and the considerations that went into his decision-making as he navigated the grant application process:

> JOSEPH: I did the application from the phone, and they sent me a letter, and after that—I don't remember exactly how many days, or week, I got help from FEMA.
> SM: Was that enough to take care of your house?
> JOSEPH: No, not really because the contractor I see for the house—they estimate about $12,000 to take care of inside for me. The car, FEMA is not responsible for that. They said I have to take a loan, and I don't want to take a loan. The reason I don't want to take a loan, when I was in school I took out a loan, but I struggled to pay that. I don't want to come back to loan again to pay for it. I don't want to do that. They send me a package. SBA sent me a package to take a loan out about $46,000. I won't take it because I'm close to my retirement, so I don't want after my retirement I have to pay a loan in my retirement. I don't want to do that. After the storm I called FEMA for an appeal over the phone. I did the application for that. They sent me a letter, and after that they sent me help for that. But it's not enough. The help they give me is not enough to take care of everything, because I got my car that was flooded, that's underwater. I have to take care of inside, and the contractor asked me about $12,000 to do that. The help I get from FEMA is not enough, but they say that I can get a loan from SBA. SBA is a small business administration. SBA sent me a package for a loan of $46,000, but I don't want to take this loan because I'm close to my retirement. I don't want to, after my retirement, to have a loan to pay for my retirement. That's why I don't think I will take it.

> SM: How much did FEMA give you? What percentage of your cost did they give you for the basement?
>
> JOSEPH: How much did FEMA give me?
>
> SM: Mm-hmm.
>
> JOSEPH: It's about close to $9,000.
>
> SM: So, you still need another around $3,000 for home repair?
>
> JOSEPH: Yeah. I feel like this amount they said I take too long to do that. Sixty days past, and when I appealed that FEMA decision to give me this amount, they said that I have to contact the SBA for a loan to do that.
>
> SM: Did you know that you had to appeal right away, or not?
>
> JOSEPH: No, didn't know. I didn't know.
>
> SM: Who told you that you could appeal?
>
> JOSEPH: Nobody tell me, I just sent a letter to tell FEMA that I have to spend more.

Those disaster survivors I talked to in Canarsie who did receive grant money on the initial application, found that the grant money was insufficient to complete their repairs. When this occurred disaster survivors had to borrow from friends, rely on adult children, and use credit cards. This was the case with Greg, a 52-year-old homeowner disaster survivor. He says that he initially received $585.00 when he first applied. Then once he discovered that his floor was collapsing, he appealed, but only received about $2000.00. He explained that the reason he received this low amount was because his repairs were urgent, and he had gutted and hauled off the evidence of his damage before the FEMA adjuster was able to come back out. Greg estimated that he spent a total of $9000.00 for repairs and materials, beyond the gutting and replacement of the water heater and boiler installed by Rapid Repairs. We discuss his experience navigating seeking disaster assistance and repairing the damage to his basement. I ask Greg:

> SM: How long after the storm did you hear about FEMA, and how did you hear about them?
>
> GREG: I think it was right after the storm. Then I would say that somebody was telling us about FEMA.
>
> SM: One of your neighbors?
>
> GREG: Yeah. First of all, we called them. Then they came over to visit and they visited the house and then they must have given us $585, for all the damage.
>
> SM: Really?
>
> GREG: Yeah, But then we take it, we don't say anything.
>
> SM: Did you all have insurance?
>
> GREG: No, we didn't have insurance and we saw that, as a matter of fact, at the

bottom of the floor, there was wood and then now on top of the wood with the pipe, after then the side of the window, and then we were talking, and we were walking on the floor now and then we hear the sinking down.

SM: Wow. That's after the FEMA guy left?

GREG: They left. Then we find out after. "What do we have to do now?

SM: Did you appeal?

GREG: No, because it was urgent, immediately what we had to do was call three guys that we saw in the street and we have to remove all the wood and all the tiles, everything, and then it was piled up outside. Then we moved cabinets and everything like that. Door, clothes, everything like that, we have to remove everything and put them on the sidewalk. Then we saw two sanitation guys and they come in and they say if we want to remove those things. We say, "Yes, we want to remove them." They move them for us, which was very nice.

SM: Did you take any pictures of the damage?

GREG: No, no, they will never take any picture. My wife was there too, and my wife asked them to remove part of the wall for us, all over, all over.

GREG: Because there was a--

SM: The mold.

GREG: Yeah, the wall was damaged. After that, then the FEMA guy came in and then he checked and checked again. After that one day then they send us about $2000 something and then we take it for the floor.

SM: Was that enough to cover the floor, or was it much more?

GREG: We had to spend some money to lift the floor. If you were to see the money that we spent at the Home Depot. That was a lot more money to spend there.

SM: About how much would you say that your material cost was?

GREG: I would say $5,000. So, the money that FEMA gave was not enough.

Another issue was that the majority of homes of the survivors I spoke to did not have flood insurance policies. There was also the surprise and confusion among many disaster survivors I spoke to who were in utter disbelief that home insurance providers could simply decide not to cover their damages on the basis that the damage was due to flooding. Not being clear about the distinction between home insurance and flood insurance also meant that many Canarsie disaster survivors inadvertently indicated on their FEMA application that they have flood insurance. Providing this incorrect information also triggered being rerouted to their home insurance providers.

Naquita, a Canarsie disaster survivor with a master's degree in organizational psychology, agreed to talk with me once she finished talking to the FEMA disaster responder who was looking into where she was with her appeal. She was quite frustrated with the process, which she began the day after Sandy's landfall, including the fact that the FEMA grant-seeking process directed her to her insurance company. She was also dissatisfied that the grant she had received was inadequate and did not consider her family situation. I asked her about her insurance coverage.

> NAQUITA: Nothing. We have home insurance, but we live in Brooklyn, and who knew we needed to have flood insurance, and we didn't have flood insurance, not to mention my insurance company is based in Manhattan, and they were closed for a really long time, and when you call them it's like, 'Don't leave a claim on this phone.' Since then, we have switched to another insurance company. However, they came out—maybe two weeks ago and that inspector basically went 'I can't believe that'—because we also had sewer backup which we had to get sewer cleaned also. 'I can't believe that the sewer backed up that high' because he sees the watermarks on the wall. And I'm like, 'You have to be empathetic. If you're in that line of work, it's not for you to judge and tell me what you believe and what you don't believe. Write it up if that's what you think, then write it up and send it in. I don't know what to tell you.' He didn't check the roof. Um, nothing was really done. He just kept saying, 'I can't believe that there's all this damage.' Well, I wish he was here from the beginning. I have a video. I have pictures—FEMA didn't want to see it. The insurance adjuster didn't want to see it either.
>
> SM: Right.
>
> NAQUITA: So, I'm just saving it on my phone until somebody wants to see what it really looks like.
>
> SM: Yeah, I'd like to.
>
> NAQUITA: And, honestly, I feel that us cleaning it out took away from the real devastation because we had to. It was mold, it's disgusting. The sewer water that was in there, the salt water. I'm like, 'You can't live like that.'

Naquita expresses her and her mother's frustrations about the arbitrariness of the assessments of adjusters that led to differences in grant awards to survivors:

> NAQUITA: I think she's frustrated. Everyone is just frustrated at this point because you figure that it's almost two months or so after the storm, and you're still trying to get things back together.
>
> SM: So, when it just happened, whose responsibility did you think it was to get things going again?
>
> NAQUITA: I guess partially ours. This is our home; we want to be comfortable.

We still have to live here. So as far as the initial cleanup and so forth—but, I mean, there's only so much we can do, and we don't have the finances to get things back together down there either, so to know that you can apply for FEMA and get a grant, that was so—I was happy. It was just like 'Oh, this is a good thing. Wow.' But now that I see—I feel like, okay, I have to go through all this extra stuff to get the help, which drives me—I'm upset.
SM: Right.
NAQUITA: I'm trying not to be mad with the representatives that's here, that's just doing their job to help me, but at the same time, I'm annoyed.
SM: Right.
NAQUITA: Because I know people that didn't have to do this, and off the top, they got all of this money, and I'm like...
SM: It can't be that one inspector is seeing it this way, and another inspector is seeing it some way else. That's incredible to me.

The Canarsie Sandy survivors to whom I spoke evaluate their FEMA adjusters' appraisals and interactions as arbitrary, lacking empathy, and working with little oversight. A recurrent theme was the concern that their damages were undervalued when compared to neighbors and also based on estimates from contractors. Naquita points to a surprising level of scrutiny and disregard for losses, respectively.

These adjusters are subcontractors hired by FEMA after a brief training and certification, which leaves little room for oversight. They are also carrying out a different role than my research participants were aware of. A 2015 Government Accountability Office (GAO) report acknowledges that FEMA hired these adjusters "to verify the identity and residency of applicants and that reported damage was a result of Hurricane Sandy" (GAO 2015, 9).

Beyond the goal of more efficient coordination, FEMA's post-Sandy reform had also involved a focus on reducing fraud. FEMA achieved this goal with precision. A 2015 GAO report commends FEMA for reducing disaster relief fraud to "2.7 percent of that total that was at risk of being improper or fraudulent compared to 10 to 22 percent of similar assistance provided for Hurricanes Katrina and Rita" (GAO 2015, 9). The report also cautions that FEMA lacked efficiency in detecting fraud by not effectively verifying Social Security numbers with the Social Security Administration. The Canarsie disaster survivors were not assessing the performance of these adjusters through this lens.

In the above sections, I use adjectives such as "FEMA-run" or "NYS-run," and later "NGO-run" centers, to remind the reader of the de facto relative, operational dominance of managers of various types of organizations at a particular center. It is important to note, however, that FEMA was present only on invitation to assist with the local efforts of disaster response. My use of "disaster response center" also encompasses "relief" "recovery" or "restoration" centers. Furthermore, the centers themselves are non-permanent assemblages of relief organizations, governmental and nongovernmental, responding to the needs of disaster survivors inside of local venues such as churches.

CHAPTER 5

Social Capital in Crisis

> Basically, the local community ran it all before.... When people came here it was a place of warming. Just knowing there was someone around, especially someone that was from the community. That was lending a hand to them despite the trauma they went through. It heals what they might have been going through.... People to talk to.
>
> —Merissa, thirty, Eastviller, local volunteer, unaffiliated

In this chapter, I discuss how social capital, the informational and resource value of informal social ties, becomes nonfunctional and inaccessible. My conversations with the local disaster survivors and local volunteers in Westville and Eastville illuminate what happens to social capital during a disaster. However, I argue that when the utility of traditional stores of social capital gets lost, there is an emergence of what I am calling "crisis capital" in disaster-impacted communities.

How Eastvillers Lose Pre-disaster Social Capital

Disasters affect the access and mobilization of social capital that is typically available in routine life. Several racially minoritized disaster survivors whom I spoke with or learned of through neighbors or responders had experienced displacement due to Sandy. In Eastville many economically deprived Black men lived in single-room occupancies (SROs). Several of these men lived in three-quarter houses or outpatient clinics in the heart of Eastville. Sandy displaced and dislodged them from their social connections. Jordan, a six-foot, three-inch forty-year-old outpatient resident worker, who also survived the storm surge, talked with me as he dragged a soaked mattress out of a displaced resident's room. Jordan explained that as an in-residence clinic worker, he "basically helped residents with their drug addiction and help[ed] with whatever they needed." When I asked what happened to the former residents as a result of the storm, Jordan answered, "Some of them are at shelters, at friends' houses, some are on the train. Homeless. Some of them are homeless!"

Jordan also mentioned many of the former residents came back to seek him after the storm. They had asked him for information and advice about seeking disaster assistance. He told me he gave them information to apply for various disaster relief and recovery programs. However, on my later visits to Eastville, the building was tagged, condemned, and vacant. Jordan would no longer be a contact for these men who may have wished to reconnect with him. Ricky, who was a former resident from the three-quarter house across the street and was now homeless, told me that many other residents from other buildings in the area were displaced and moved to hotels in an entirely different borough. Unknown to him at the time, several months later he, too, would follow.

Another compromise to racially minoritized disaster survivors' access to pre-disaster social capital was social network resource deflation. This deflation of network resources was due to the stripping away of social resources from long-standing social ties. The social networks of the economically deprived and racially minoritized were concentrated within the same apartment, building, or block. This meant that social ties were equally affected by the storm surge and could not provide housing, monetary, or food assistance. This stripping away of meager resources held by these social ties made them unavailable for providing disaster support.

Among the few who had familial ties outside the neighborhood, some mentioned that these once close ties were severed even before the storm. For example, Ricky mentioned that he and his sister, who lived in New Jersey, were estranged and that he would not ask for her help. He explained that she had a drug problem, and since he had been clean for seven months, he was not "tryin' to do that." Others talked about family rifts that could not be mended despite their predicament.

Even for those with friends in other states, the resources of these ties still proved unusable several months after Sandy. Rose, a sixty-three-year-old African American disabled former lawyer with ties in California, explained this to me:

> My birthday was Christmas, and well-meaning people were actually sending me checks.... and I said, "There's no bank. Where could I possibly put this?" They say "Well, is there a check cashing?" I say, "Dear, there's no electricity. [*Laughs*]

Even in cases when economically deprived residents had unimpacted social ties within a reasonable distance, they quickly exhausted the resource-conferring capacity of these social ties. George, an economically deprived Italian SRO resident, explains why he returned to his room despite having no heat or electricity:

> Oh yes, I've been to the shelter. But how long can you stay in a shelter? [*Spreading his hands out, shrugging his shoulders*] I've been with my niece for three weeks. You know, she doesn't want me there no more. [*Shrugging his shoulders, indicating that this was also reasonable*]

Several racially minoritized and economically deprived Eastville and Canarsie disaster survivors who stayed with family or friends initially moved out from these homes only to sleep in cars, return to extremely cold and dark rooms, or move from shelter to shelter. Harold and Henrietta, a retired sixty-nine-year-old and seventy-three-year-old Black married couple living on a fixed income, discussed their multiple displacements and moves induced by the storm. Harold, who is from Jamaica, explained:

> That time, when the hurricane hit, I wasn't with my wife. My wife said I must come with her [to her daughter's house]. I said, "No, I'm not coming." So, I went down by my sister and stayed there, and then the water started coming into the basement. Then from there, we started to bail the water out, and the hurricane started, and then when the hurricane passed through, then the water started to dry out. Don't see no more water. Then I went down by my house. Yes, when I went back home, the first thing I opened the door, and I looked in there and I said, "Oh my God. Somebody was in here." The fridge was on the back. The bed was flipped around like that. When I looked in the next room it was full of water. I came home and went to my wife and that was it. Then, without nowhere to sleep we have to sleep in the van one night and like the gas was bad too. We weren't getting any gas, so we went by the gas station in the line and sat in the line till I got gas, and by the time I got gas went back home. From home to the hotel, from the hotel back home. From the hotel you go back home; from home, we are over [at the Westville center] now. So, it's not really saying, it's not one place. We went back and forth, back and forth. It's a good thing I got the truck too, like when I tell my wife, "I'm not coming." She says, "Why you not coming?" I say, "No Honey, I'm not coming. You go." And that's the only way I could save that truck from water flowing. When I park it at my sister, look through the windows and see water start to come up. I run out and I move it to the next sister house, that's how I saved that van up till now. Still running.

Beyond these structural and functional issues with informal networks, there are also race and class rules around asking that complicate mobilization of pre-disaster social capital of survivors. Despite the sociological literature's emphasis on the social support value of kin ties among Black families (Stack 1974), there are also cultural rules about asking for help, regarding when one should ask, what can one legitimately ask for, what should not be asked, and limits as to when one has expended their acceptable amount of support (Hansen 2011).

All of these asking rules constrain the social capital value of these ties. Studies on social class have also shown that working-class individuals are not as inclined to ask for help as are the middle class (Calarco 2011). In the above example with Harold and Henrietta, despite having extended family in the area on both sides, this couple still resorted to sleeping in their van in extremely wintry weather, incurred multiple "back and forth" trips up to seven times between their home, family, van, shelter, and hotel. Even as I spoke to them that day, they were still contemplating where they would spend the night.

How Westvillers Lose Pre-disaster Social Capital

Sandy also compromised the social networks of White economically privileged Westvillers, but in ways that differed from Eastvillers. The geographic extent of the networks of White economically privileged disaster survivors meant that their social ties were typically outside the impacted area and often stretched into other affluent neighborhoods. One superintendent of a cooperative building explained, as we walked through a gutted-out, first-floor apartment, that residents had dispersed throughout Long Island, New Jersey, and even to Florida. Such hiatuses were an avenue for some to "escape" for a while, but they eventually had to come back and face the devastation.

However, those who lived in co-ops were the fortunate ones because they were able to use their social capital outside since they didn't have to be present to do or oversee repairs to their homes. However, for the Westvillers who were "stayers," most of them were homeowners of single-family homes, which meant that they were forced to remain in Westville to repair and protect their homes from further damage. This made their nonlocal social ties unusable for the purpose of providing shelter. Similarly, the storm also destroyed places of gathering, communication technologies, and transportation channels, also making most organizational ties of White economically privileged Westvillers inaccessible.

Another form of social capital that White economically privileged Westvillers had enjoyed prior to the storm was their ability to gain favors based on their reputation or status due to being board members or having a personal connection to a local bank. However, Sandy had created a situation where demands on social ties, coupled with elevated transactional risk, far exceeded the ability of social ties to confer preference. Due to the overload of requests from several disaster survivors, favors were not as forthcoming. In fact, several landlords and well-to-do disaster survivors expressed frustration with not being able to access this form of social capital.

Furthermore, disaster survivors had needs that were different from what they would have been under regular circumstances. When I asked Joe, a sixty-year-old White Westville disaster survivor and landlord with tenants in Eastville, what he found most stressful, he explained:

> Well, nobody was really prepared or had any idea of how to deal with all this—It's like everyone in the community all of a sudden had to pump out their basements, try to deal with mold, do demolition, try, and—try to reach out to plumbers, electricians to put their homes back together again. We're all used to getting things immediately! You make a phone call, and someone comes out the next day. So that was stressful.

Similarly, Peter Dexter, an eighty-year-old small business owner whose establishment was directly across from a bank he patronized for thirty years, expressed his disbelief and frustration that the local bank, which had now partially resumed operations, would not approve his business loan application. He could not accept that he would not get preferential consideration based on his reputation of successfully running his business for thirty years directly across from and having a close relationship with the bank. Dexter was surprised to learn that his credit worthiness would now objectively be based on the current state of his damaged building and neighborhood.

Other White affluent Westvillers enjoyed a more generalized form of social capital, in the form of preferential treatment that hinged not on personal reputation but on status (Smith 2005). In this situation of crisis, when the number of residents who needed favors exceeded the ability to deliver, it rendered this form of social capital unavailable. One more available form of social capital after the disaster was the access to organizational social capital through church membership. Resident parishioners went to their churches for help and also shared this information with their neighbors. However, since church resources became a public good, these benefits were extended beyond supporting parishioners.

The most vivid example of sustained pre-disaster social capital I observed was between Sylvester, the owner of a print shop, and Johnny, a loyal patron. I stood interviewing Sylvester, who was a visibly despondent owner of this print shop, which reeked of backed-up sewage. On the walls hung shelves of paper and other unrecognizable things that were waterlogged, moldy, and damp. There was debris everywhere. However, Johnny, a disaster survivor, walked in with a smile, called Sylvester by his first name, and placed an order for a print job, pretending like the store was still how it looked before Sandy. This was a gesture of encouragement to the business owner, but few patrons had that much confidence in their own ability to pay for or use such an order.

Crisis Capital Emergence in Urban Disaster Areas

In the context of the unavailability of pre-disaster social capital during the early disaster response period, crisis capital becomes significant. Crisis capital is a transient form of social capital, characterized by "warmth" in addition to support, which stems from the local community (i.e., residents, grassroots organizing, community-based organizations). This crisis capital is quintessential to disaster response efforts even before nonlocal responders arrive with their organizations. Crisis capital is crucial to the survival of economically deprived urban communities, but the arrival of a large NGO interrupts the emergence of this significant form of local disaster social capital. Large nonlocal NGOs have access to sustainable streams of disaster resources, but they do not pursue amicable relations and exchange with local volunteers, grass-roots efforts, and community-based organizations. Yet, amicable relations with economically deprived disaster areas, if pursued, would augment and support the crisis capital in these communities.

Long before Resiliency Is Us and other nonlocal NGOs arrived, organizations and persons within those communities actualized community crisis capital. In an ideal situation, crisis capital would serve as a bridge between the loss of traditional stores of social capital and a new social capital connected to a steady stream of institutionalized resources.

Crisis capital is different from accessing preexisting social capital; rather, it actualizes, or sets in motion, the social capital potential of communities. Community disaster survivors with no personal connections or associational memberships before the disaster primarily relied on crisis capital. Even after the disaster, crisis capital, unlike traditional social capital, did not depend on having a specific relationship or tie among actors.

However, if I were to conceptualize the existence of a social tie, it would be a generalized connection between the actualized actor and a class of people or a specific demographic group such as "the elderly" or "the poor." In this case, actualized crisis capital does not direct to any specific person but is available to anyone who is a member of the specific category. The last distinguishing quality of crisis capital is that once formed, it becomes a public good available to everyone in that class, such as "Sandy survivors."

The source of actualized crisis capital is usually from residents and persons living in the disaster-impacted community or who have a direct connection to someone in that community. Typically, these higher-resource actors have already gone through the disaster experience themselves. This type of local transient social capital is vital within the first few days of the disaster, even before the official first responders, police, fire, and sanitation workers get to survey the damage.

In economically deprived communities, such as Eastville, this may be the only or the main source of capital available for a long time. For example, when speaking to Merissa, I asked about what motivated her to give disaster help to her community—in other words, to become a transmitter of crisis capital. Merissa explained:

> Even though I was impacted, the only thing was loss of water and loss of light, but I still had a home to go to sleep. And I never thought that when I walked through this door someone would say, "I need a hand. Can you help me provide services for residents?"

The true resource of crisis capital is "know-how." This resource is not present prior to the disaster and actualizes as a result of the disaster experience. In other words, it is not preexisting. Also, the resources from these actualized social ties are finite, and so the value of this form of capital is inherently unsustainable. One source of actualized community crisis capital was disaster survivors who began baking cakes and brought them to the church before Resiliency Is Us came, until the monsignor deemed it unsanitary.

Similarly, Rudy, a fifty-six-year-old Westviller, describes how the owner of the grocery store "emerged as a leader" in organizing the community disaster response efforts and was doing well, but then he got tired and frustrated and left. As these actualized ties first became inundated and fatigued and watched their resources deplete, their capacity to assist also faded. Another form of actualized community crisis capital came from former residents, such as Kacie, who came to volunteer in Westville because she grew up in Westville. She had a sentimental connection to the "memory" of the community as a child, but no personal or associational connections left there.

Yet another form of actualized crisis capital is the grassroots community organizing in order to provide disaster response. This type of capital is therapeutic and well received by residents, but fleeting. For example, Rudy describes his connection to a local grassroots organization. "People bonded with Apple Angels people because they gave us food and they took care of us. They're not considered first responders, but they were the ones who fed everybody.... It's the energy they gave out that had an impact." After the out-of-town NGO took over from these local organizations, survivors felt the loss. Rudy says that his dog misses them because "she bonded with the people here with Apple Angels. Everybody knows her." The greatest value of this form of capital stems from the intimate knowledge of the community needs that enable customized assistance to be deployed quickly, particularly to less fortunate and hard-to-reach demographics. Despite the utility of this form of capital, like other forms of actualized crisis capital, it is often short-lived due to fatigue, overload, and resource depletion, making it unsustainable for the duration of a protracted crisis and recovery.

The Importance of Local Altruism to Eastvillers

Local community volunteers and community-based organizations such as Always With You know of chronic needs in their communities, so it is not necessary to extensively communicate those needs to them. When I asked Merissa whether she had talked to disaster survivors who came to the make-shift structure before Resiliency Is Us set up its operations, she replied, "Yes I have." Then she immediately interrupted herself. "We had problems before Sandy came," revealing that her awareness and understanding of needs came from her own local knowledge of persistent issues. She then gave an example of one such problem. "Senior services people could not get out. Elevators did not work. One building has eleven floors." Merissa exercised agency on behalf of her community. She told me: "I had raised the issue about sending people over for the senior citizens. There were responses. People were going to each door knocking on the door from top to bottom. I believe it was Apple Angels." I asked, "You suggested that to them because you knew?" Merissa answered, "Yes, I knew of the situation. Then the media put pressure on things."

This type of assistance, in the absence of solicitation, provided by community volunteers reveals an intimate knowledge of communities. This in-depth knowledge of the specific type of vulnerability residents would be experiencing is indispensable in providing disaster assistance. As was true for Merissa, the ability to empathize partially stems from similar experiences before and during the disaster.

Community volunteers are also uniquely positioned to work collaboratively with local community-based organizations, creating a more synergistic response than nonlocal NGOs could. Community-based organizations founded by community members have continually had to respond to chronic crises of unemployment, poverty, racism, illegality, drug dependency, overpolicing, diabetes, mental health, and so much more in their communities. Always With You was already actively serving the local Eastville community in that capacity. This means that their relationship with community members was intimate and well established before Sandy. This led to a holistic responsiveness that caters to a range of needs. Sapphire discussed an example with me:

SAPPHIRE: There's a big family from El Salvador. Three of the sisters lived here... two sisters who lived there with children [*pointing to a vacant lot*]... had to run out of their homes in the flood to escape the fire.... There were seventeen people living in a one-bedroom home. You know?... we're talking about very unacculturated El Salvadoran immigrants [who are] day laborers, informal workers that have had very little access to education, [who] barely speak the language in terms of knowing what aid was available for them, who they can trust, who they can reach out to, what services they were eligible for. And solving the most basic problems became twice or three times more complicated for them than anyone else because of lack of access to language, cultural aspects, and just because they lost everything... with the family dealing with loss.

SM: How did they hear about you?

SAPPHIRE: They came to Always With You from the first day asking for donations. I took it upon myself to be that person to work with them since I was one of the few volunteers who was here who spoke Spanish, so I ended up meeting all these families. The first day we provided them with clothes, blankets, flashlights, but eventually medical services. At one point everyone in the family was sick. We were able to bring in a doctor to evaluate everyone in the home and provided antibiotics to a couple of them. We also eventually, now that they are finally in the fourth month, have new apartments to move into. We try to get them furniture and beds. You know, we've helped them out with FEMA applications and their appeals.

Here we see the seamless response to cascading chronic and acute crises of survivors. Community-based organization volunteers and staff are already adept at working on an amalgamation of issues when dealing with the economically deprived urban residents. Furthermore, the people who are attracted to work with these organizations already understand these demands.

Unsolicited Assistance and Reciprocal Loyalty

Always With You responders and Eastvillers had strong bonds between them. However, Always With You did not have the direct and sustained access to resources from large corporations, donors, and the State that Resiliency Is Us does. Always With You's founder and volunteers demonstrated strong commitment to the community beyond the disaster as they were aware of the chronic problems with the community residents.

This means that disaster survivors can automatically benefit from *actualized* social ties, without explicitly making a case for their needs. Always With You responders understood what those needs were. Eastvillers, in turn, repaid the psychosocial relations with loyalty and confidence. This loyalty is similar to the loyalty one would expect of regular patrons to their barber or hairdresser. You walk in, sit down, and they know what you need without explanation.

Community members established pre-disaster bonds with the Always With You founder and volunteers. This is evident in my discussion with Eric, a forty-six-year-old Native American:

> SM: Who has reached out to you and helped you?
>
> ERIC: Most certainly all churches have stepped in, Always With You is another group, the Occupy Sandy people have helped feed us and, uh, get us assistance, and, uh, Resiliency Is Us has shown up. Maybe five or six days after the storm they were here.

Eric mentions Always With You, local churches, and other grassroots movements as helping and Resiliency Is Us as merely showing up late. Months earlier while sucking deeply on a cigarette and shivering from the harsh winter's sea breeze, referring to the make-shift structure where Resiliency Is Us would set up their disaster response center, Eric had told me: "There is a disaster response center down by the Westville Catholic Church. From the first night people have brought in clothes. See? [*He pinches at his winter jacket*] I'm wearing a nice warm jacket." [*Smiles*] Months later, standing inside the warm Always With You center with broom in hand, he demonstrates his loyalty and reciprocal commitment to Always With You, stating, "I'm here trying to sweep up for Always With You and helping out and giving a little bit of my time. Always With You is one of the first groups that came and fed us. They gave us hot meals, so it is a way to return that favor."

Eric immediately follows with commendation for the founder of Always With You: "Freddie is a great person. He's done a lot of stuff in the community," revealing the broader community context he is using to assess Always With You's worthiness of his patronizing the center.

How Resiliency Is Us Interrupts Eastvillers' Crisis Capital

Resiliency Is Us did not collaborate very well with Always With You in Eastville. Eastvillers perceived Resiliency Is Us responders as nonlocals who came to "take over" daily operations of local grassroots disaster response efforts. Merissa was the only remaining local volunteer at the Resiliency Is Us Eastville location. Her tenure began even before the first responders and Resiliency Is Us arrived. When I inquired about the notable absence of Eastvillers, Merissa explained that ever since Resiliency Is Us took over from a local grassroots organization, it ushered in a more "organized" disaster response effort, which made supplies inaccessible to the community and drained the community of "warmth." This action by the NGO resulted in Eastvillers' lack of participation in the Resiliency Is Us disaster response center. Merissa explained this situation to me:

> MERISSA: The day after the storm when the water receded, we had Apple Angels [a local community-based organization] out, but we also had local churches that were also giving a hand. There was no place to place things inside a building. They were just placed on sidewalks or steps of religious organizations. [This] church was accepting a lot of donations. It was a large space where people just dumped their items here, and someone came and thought of putting things in this make-shift structure. Now it became organized with donations and distribution in one structure and food in one. With everything being so organized [*sarcasm*], a lot of things people need—they're being placed somewhere else. Resiliency Is Us has a permit, but basically the local community ran it all before. Resiliency Is Us came in about late November [four weeks after the storm]. Most of it was food in the trucks. Local persons were running the tent.
>
> SM: What difference did you observe with the transition?
>
> MERISSA: Well, when people came here it was a place of warming. Just knowing there was someone around, especially someone that was from the community, that was lending a hand to them despite the trauma they went through. It heals what they might have been going through. Because they had someone from the community that they were familiar with to give them a hand, feed them, and give them supplies. Things that they needed as well as having some form of mental consultation. People to talk to. So, it's basically in spite of what they had lost, they had somewhere to come to, and they felt comfort coming to this location.

What Merissa was describing was what Rebecca Solnit (2010) describes as an altruistic utopia. Solnit (2010) explained Mizpah Cafés as egalitarian kitchens created and run by local volunteers after the 1906 San Francisco earthquake. Merissa's utopia was the coalescence of a local community, local community-based organizations, and local volunteers who created a "place of warming" and "healing" where disaster survivors had access to needed supplies. This utopia also provided "mental consultations" and "people to talk to." But unlike Solnit's post-disaster utopia, Merissa describes the abrupt interruption of this organic emergence of community crisis capital triggered by Resiliency Is Us, the nonlocal, large NGO that simply came and "took over." This utopia was no more, as I stated to Merissa:

> SM: It looks pretty empty.
>
> MERISSA: Yes it's pretty empty. We have been going through a lot of changes. Who is supposed to take charge? The main thing it boils down to is bringing back the communities together and having them come back to where they felt warmth. So, it's not just that the space was warm. They felt warm in their hearts.

Merissa indicates that the management of disaster response by Resiliency Is Us left a situation where it was not clear who was "supposed to take charge." This power dynamic that ensued impacted the quality of the disaster assistance that Eastvillers received. These Eastvillers no longer felt the "warmth" emanating from their pre-NGO-arrival crisis capital utopia, and now they no longer significantly participated in the disaster response center.

My observations of interactions and interviews with Eastvillers confirmed Merissa's viewpoint that residents were comfortable in their relationships with local volunteers from their communities, due to the "warmth" they provided. Racially minoritized Eastvillers almost entirely disengaged from utilizing the center under Resiliency Is Us management.

CHAPTER 6

Logic of Response versus Services

> The main object of the long-term recovery group is not to make people better off than they were before the storm, but to bring back people to some stability, where they were before the storm.
>
> —Reverend Dennis, Brooklyn pastor, disaster responder

My conversations with disaster responders from both governmental and nongovernmental organizations across Brooklyn and The Rockaways revealed that there is an orientation in disaster response that only "sees" acute trauma, occluding chronic trauma from consciousness (Erickson 1976). These conversations reveal that a *logic of services* that cater to the chronically economically deprived had given way to a *logic of response* that prioritized those disaster survivors who only momentarily were without access to their personal resources. I was often amazed at the assumptions and what seemed like a lack of awareness among well-meaning disaster responders that many "disaster survivors" or "disaster victims" were in fact severely economically deprived before Sandy. These discourses and practices of disaster responders together form what I am calling logic of response versus logic of services, which prioritizes and promotes "victims over vulnerable," "middle-class bootstrap," "color and class blindness," "primacy of homeownership," and "self-employed invisibility." In particular, the "victims over vulnerable" logic of response was the most dominant discourse across two FEMA-run disaster response centers, one NGO-run center, an NYS-run center, and one long-term recovery group community meeting for local Brooklyn churches, local businesses, and nonprofits responding to the disaster.

Victims over Vulnerable

The victims over vulnerable disaster response logic reoriented definitions of need and deservingness to include anyone impacted by the storm. This tendency conflated the chronically economically deprived with the economically privileged disaster survivors whose plight was only temporary. Disaster responders use references such as "disaster survivors," "disaster victims," or "impacted residents" in discussions. This results in a linguistic displacement of the truly economically deprived. However, this displacement is more than just semantics. Claims making for securing disaster aid is also about proving that one's economic standing is the consequence of the disaster event. I illustrate this victims over vulnerable disaster logic of response in my conversation with Reverend Dennis at the end of a long-term recovery group meeting held at his Brooklyn church, with several Brooklyn area local churches and nonprofits. Reverend Dennis explains:

> REVEREND DENNIS: The main object of the long-term recovery group is not to make people better off than they were before the storm, but to bring back people to some stability, where they were before the storm.
>
> SM: But when you think about people in the neighborhood who were already in bad shape before the storm, what are the alternatives for them if you're not going to be putting them in better shape?
>
> REVEREND DENNIS: That is a question that case management has to deal with.
>
> SM: Case management. So, tell me more.
>
> REVEREND DENNIS: In fact, the case management will be able to help them assess their circumstances. See what their needs are and see where the resources to bring them back to a better position than they were before the storm can be. For example, they might be able to direct them to social services or to other services that can help them move from where they were before the storm to a better place.

Here, Reverend Dennis is demarcating a line between the concern of disaster response and that of social services. However, this line also represents a demarcation of logic of services versus logic of response, where the latter does not imagine a disaster victim as being already unstable before the disaster. There is an arbitrary line between a disaster survivor and the chronically economically deprived. This logic also surfaced in my conversation with Bob, a mental health counselor with Resiliency Is Us in The Rockaways. Bob explains the role of the NGO:

> It's important to remember that we are a disaster relief operation. So, we're not necessarily here to provide psychotherapeutic or longer-term counseling services.

> We are here to assist people in overcoming and dealing with the immediate crisis. So that they can marshal their resources—both the personal and community in order to allow *their own personal resilience* to kick in.

The statements of both Reverend Dennis and Bob illustrate what I was hearing when speaking with Always With You volunteers on The Rockaways, regarding the plight of chronically economically deprived survivors. That is, the needs of the subpopulations that are most vulnerable to the disaster event are not the business of disaster relief and recovery, but that of other social service agencies, churches, and other private organizations.

This orientation was also an outgrowth of the post-Katrina emphasis on resiliency. Reverend Dennis and I along with representatives of twenty-five or so Brooklyn organizations had just sat through a meeting where "resiliency" was the resounding theme and was discussed as a strategy to competitively position this newly forming long-term recovery group to receive Sandy relief. Attendees at the meeting of small local churches, businesses, nonprofits, and a large NGO echoed Reverend Dennis's statement, with hardly any rebuttal.

Reverend Dennis and Bob were also making assumptions about survivors' personal resources or social capital, which are important for these disaster survivors to achieve this desired "resiliency." The prevailing disaster logic of response is that the imaginary "disaster victim" is someone who is at the very least not economically deprived and is well connected or supported by a well-resourced family and community. Also, both statements illustrate that these disaster responders are not providing the kind of help that would benefit those who lack the pre-disaster resources because such help is not the mission of "long-term recovery" nor of shorter-term disaster response NGOs.

Another aspect of disaster response that reflects an orientation toward the logic of response is the spatial organization and operations of FEMA-run disaster response centers. These centers can become a revolving door for disaster survivors needing critical services. The physical layout of these centers resembled a job fair setup, outfitted with rows of staffed tables with small signs in a large vacant building or church. If a disaster survivor needs services outside of what these tables offer, there are flyers and brochures with the appropriate agency to contact for their specific need, but the onus is on them to call and make contact.

The design of these centers resembled a one-stop shop. However, the experience for the disaster survivors is that they go in to see one state governmental organization represented at a table, only to be sent to another table, and another, and another. Each successive table is closer to the exit door. For those who do not receive the needed assistance that they are seeking from any of these tables, they make it to the exit door and onto the streets with unmet needs and are visibly frustrated, distressed, and despondent.

This was the case with Marlene, the seventy-three-year-old retired seamstress and Canarsie homeowner on a fixed income from Jamaica. I met Marlene when she was leaving the FEMA-run disaster response center, with a look of frustration. I introduced myself, told her about my research, and asked if she had a few minutes to participate in an interview. During the course of the interview, she broke down crying profusely. She began to say that she woke up that morning knowing that she was going to jump off the Manhattan Bridge if FEMA did not help her that day. FEMA did not help her that day. I immediately turned off and put down my recording device to empathize with her and to dissuade her from these suicidal thoughts. Surprisingly, she said, "No, no, I want you to turn it on. I want everyone to know what has happened to me." At that moment, I was torn about whether to keep recording, but she affirmed that the very process of telling me her experience was cathartic and an empowering experience for her.

Marlene relayed that she was seventy-three years old and that she had become temporarily homeless because her basement, where she lived, was flooded. She said that all she was walking around with was her bag, which she revealed only had two pairs of "panties" or underwear. She said that the night before she had slept at a friend's apartment near Ocean Avenue in Brooklyn and was not sure where she was going to sleep that evening since she had been displaced from the basement, but she rents out the upper floor of the house to be able to pay her mortgage.

Marlene had made several trips to FEMA and had been directed to the tables represented by FEMA, NYS, SBA, HUD, and more at the disaster response center. Yet, she felt that all these bureaucracies that were set up to help her did not really hear her pleas for help. Recording the interview validated her experience and her way of creating recorded history as if to say, "Yes, this really did happen, and it happened to me."

Because Marlene had revealed she was having thoughts of taking her life, I sought assistance from one of the tables and was given the crisis hotline number. I then called the crisis hotline to connect Marlene to mental health services using my cell phone. I spent forty-five minutes listening to several automated options, bumped from one menu to another before getting a live person on the phone. I was finally able to give Marlene an address to go to, based on what the operator relayed to me.

I called Marlene the next day to see whether she had received assistance. However, she told me that she had taken the bus to the address, but when she arrived, she discovered that it was an abandoned building. The information was obsolete. I went through the entire process again before she could be seen. Marlene also gave me permission to share her needs with other community disaster responders in another area, and I was able to find a team to help her with mold remediation.

Before this, Marlene had been trying to battle mold in her basement by herself by using bleach she bought from Home Depot, only to find that the mold grew right back. Since she had communicated her mental state, I wanted to minimize her stress as much as I could. It was at this moment that I realized that this disaster response bureaucracy was a huge referral system that did not allow the time and opportunity to really "see" and meaningfully assist people in Marlene's situation. It was just too easy to slip into the chasm between the logic of response and the logic of services.

There is a fundamental problem with establishing a bright line between a logic of response and a logic of services. It is equally problematic when the former displaces the latter. The reason is that the chronically economically deprived who rely on social services are also those who are most socially vulnerable to disasters. This means that when these disaster survivors walk into a disaster response center for assistance along with everyone else, they cannot disentangle their needs along a false dichotomy.

My conversation with Caroline, a site manager of an NYS-run response center, illustrates how from the earliest period of official disaster response, the logic of response displaces the logic of services. This displacement of logic relegates the latter to the position of an afterthought. Caroline's account also sheds light on the tangible, negative impact of disaster response in the absence of a thorough integration with social services. Prior to managing the NYS-run disaster response center, Caroline had been stationed at an evacuation center. She laments over her own observations at the evacuation center regarding the "areas of service that were missed."

> Well, it was, uh, it was emergency management, um, that really needed to link up with social services. Whoever came up with the plan just didn't link us. And so, we were all there, but there was just no linkage. And so, the people were brought out of the storm. There was transportation, which was fine. It was hot, and it was warm, you know. It was safe. There were cots, but there were all those other service opportunities that I think, um, could have been addressed so much more systematically as opposed to what's going on. Like, "Oh, my gosh. The people need that." And then doing it informally. I think that, systematically, we'll probably learn from our experience with Sandy and be better prepared in a different way, not just with the stuff, but with the, you know, with the systems in place to do a better job the next time.

Caroline's vision of a systematic integration of the logic of services with the logic of response was different from that of the disaster response site managers with whom I spoke at the FEMA-run centers. Her vision of integration shaped many aspects of the way she managed her disaster center. However, in order to achieve this, she constantly had to advocate for more services for the disaster survivors coming into her center.

> I'm always thinking about the agency, what we should be doing. You know, I'm in this project now, so this is my agency, you know, and it's important to me that it works as effectively as it possibly can. And so, either they love me or hate me because I've always got an idea about what would make it better, because I sit here and I watch, and I don't think that, with all the money that was being put into making this work, anybody should miss services or not get what they need, because nobody said anything.

Caroline was indeed an anomaly. To offer some context, Caroline is a licensed social worker who has worked with the social service agency for over thirty years. She told me, "Social services is kind of in my blood." She explains that her own mother was a social worker who worked with the agency. She acknowledges that exposure to this world from an early age has profoundly shaped her worldview. She has also worked in a homeless relocation program that connected these clients with services and local community-based organizations around their new home. She laments the decision to end that program.

Unlike the other two site managers in the other FEMA-run centers, Caroline brought in a community-based organization to provide on-the-spot counseling beyond the regular referral system. Her team was racially diverse and mostly local. My conversation with her also reflected that she was culturally aware when she noted that Caribbean, African American, and Latinx disaster survivors don't tend to ask for mental health services. Then she demonstrated her responsiveness by identifying them and connecting them with the therapist, even closing off an area from the open warehouse setup of the center.

Caroline's integrated approach is a great model for disaster response. At the institutional level, NYS-run disaster response centers are more equipped to effectively serve the local community. Disaster response decision makers need to select site managers from a pool of local applicants with a similar record of accomplishment in social services. Leaders with a social services perspective are ideal for informing a disaster response model that is more responsive to the needs of the chronically economically deprived and marginalized urban disaster survivors.

Middle-Class Bootstrap Bias

In addition to their displacing the economically deprived, the elderly, and those needing long-term mental health services, there was an assumption among governmental disaster responders that the ideal disaster survivors were those who were the early comers pulling themselves up by their own bootstraps.

Middle-class bootstrap bias surfaced in my conversations with FEMA staff and state field site managers across Eastville, Westville, and Canarsie. Several studies have shown that institutional expectations tend to align with middle-class ideals or "cultural capital" in ways that they don't with the ideals and cultural capital of the working class (Calarco 2011). I saw and heard examples of middle-class bootstrap bias in the way responders spoke favorably about disaster survivors who they thought were being "proactive." Contrastingly, it was also apparent in their assumptions of survivors' lack of initiative as explanations for unfavorable outcomes. In some cases, these were personal ideals among responders, but often these were also woven into institutional practice independent of personal ideals as illustrated below.

Through my interviews with low SES Canarsie and Eastville survivors, who were at the lowest rung of the socioeconomic ladder, it became apparent that poorer and newer immigrants and the racially minoritized heard about programs a lot later than White middle-class residents in Westville—the more affluent community on The Rockaways. As I alternated my presence at different sites, I realized many times that the Canarsie and Eastville disaster survivors were just hearing about a program that White disaster survivors in Westville had talked to me about several weeks before. In fact, in some cases, Eastvillers and Canarsie residents were learning about programs from me.

Both in Canarsie and Eastville there was a lot of displacement of those who lived in basements and who lived in coastal areas. Displacement interrupts the flow of information because of the fragmentation of informal networks. Ferdinand at the Canarsie FEMA disaster response center mentioned that initially Haitians had low representation in the earlier flows of resident applications coming to the center. He noted that more recently a lot more had begun to come in. He attributed the recent increase to word of mouth among Haitian immigrants. Through conversations others learned that being part of a mixed immigration status household would not negatively affect the applications of legal household members.

It is also important to note that Ferdinand was also critical to this reconstituting of what later became an informal informational egocentric network that formed around him, largely because he was also Haitian. Lots of Haitian immigrants had incorrectly assumed that they did not qualify for disaster assistance, but his presence as an embedded actor at the disaster response center legitimated him as a trusted source, although he later told me this placement was quite by accident and not a matter of design since he had been scheduled for another location before this one.

This late emergence of a network is not surprising, given that we know that information tends to diffuse in social capital networks of ethnic groups. In the migration literature, Alejandro Portes (1998; Portes and Sensenbrenner 1993) confirms the importance of the social capital networks of migrants and ethnic enclaves by stressing the importance of this homophily principle on the basis of shared ethnic identity. A common identity through national origin and migration experience forges strong affect, trust, and solidarity among co-ethnics, which facilitate reciprocal exchanges of obligations and fulfilled expectations. Immigration scholars think of the social capital residing in these immigrant networks and enclaves as a response to social inequality due to discrimination in the formal labor market and exclusionary communities.

Color and Class Blindness

Other disaster responders did not possess Ferdinand's awareness that there are socially determined, external factors that may explain the lag time among some of the Canarsie survivors. Beverly, a White disaster responder under his supervision, who staffed one of the service desks with which residents interfaced, explained her work process of mold remediation as a class-blind, color-blind approach to support provision. I asked Beverly:

> SM: Can you just tell me a little bit about your experience here in terms of what you noticed about the community, the people who came, and the types of issues that you've had to help with?
>
> BEVERLY: I talk about mold issues and insurance and how to stop the mold in its tracks. That was initially. Then as it progressed it got to people who did some work and then didn't do anything or they did. It just got overwhelming.... Now what I am getting is people—it's been two weeks—who have not done anything—the basement or anything at all, have not [removed] the Sheetrock, have not kept up with the mold.
>
> SM: What sense do you get about these people? Any differences that you discern between those and those who came earlier?
>
> BEVERLY: No. Well, the ones that came earlier are a lot more *proactive*. You know, they knew they had a problem. They knew they needed to register with FEMA. They knew they needed to get some funding. Hopefully, some help

from us, and they knew that they were going to have a problem because of the type of flood water this was. It was pretty nasty water. So, they knew that. The type I'm getting now are not.

SM: So, you think that probably reflects socioeconomic status, education level or any— [*Interruption*]

BEVERLY: I'm not saying that. But then you know *I don't see that. I don't see any of that in people and I don't recognize it and I don't ask it. So, I don't see, I couldn't tell you.*

SM: Oh, so once they come to you—[*Interruption*]

BEVERLY: *They come to me just clean.* I have no information about them. When you register with FEMA over here [*pointing to the first intake table*], they take information. I don't have any access to that. Know all of our computers have different securities. Um, so I'm just dealing with them one on one.

SM: Okay.

BEVERLY: The ones that are coming in now just seem to be, um, *too relaxed*. They're not as worried about the health issues *as they should be*.

SM: So, from what I've encountered from the different people that I've spoken to, I've started to see a pattern among people who are probably with less education probably, they're just not aware [of the dangers of mold]. They just know it's an annoyance, but they're not aware that it's a health risk.

BEVERLY: See, I don't know that because I don't, like I said, I just talk one-on-one to the person. I just listen to what their dilemma is and then try to walk them through that. You know, I never ask them what they do for a living, or you know.

Unlike Ferdinand, Beverly did not consider the possibility that those coming in later were just learning of the presence of the disaster response center or their eligibility and how their social standing may have influenced their access to relevant information in a timely manner. In fact, she continued to reiterate that she was blind to socioeconomic, educational, or any other differences. As she stated, the people who came to her were "clean," so she dealt with everyone in very much the same way, considering only the needs that were induced by the disaster and nothing else.

Beverly recognized residents who came in earlier as being "proactive" and those who came in later as being "too relaxed" and not as worried as "they should be." Another favorable description of the early birds I heard from Andrew, a Black manager of a FEMA-run site at a Rockaway location: "First you get the needy, then you get the greedy." Both comments reflect a lack of awareness of the impediments to early arrival at the disaster response centers. Both statements also reflect either the survivors' need for greater personal responsibility or the undeservingness among the latecomers, common tropes used for the racially minoritized and the economically deprived. Andrew's perception that the needy would be the first to arrive runs counter to everything we know about the most marginalized and their ability to be first in line for anything, and certainly not before the most privileged.

While Beverly's and Andrew's statements reflect their own assumptions, there is also a degree of institutional culture and practice at play as well. Federal and state operations, protocol, and technologies that organized and compartmentalized social identities and prior socioeconomic status effectively strips residents of the recognition of their social vulnerabilities.

Although FEMA collected some of this personal information, the data was not immediately accessible across the different stations that residents encountered. I confirmed this by asking the workers at the various stations. At each table a resident approached, they became a clean slate and nothing more than a disaster victim. On its face this seems admirably equal, but race scholars have long argued that "color blindness" (Bonilla-Silva 1997), and in this instance I would include class blindness, actually has the opposite effect. These practices and cultures of color and class neutrality definitionally occlude the possibility of truly being able to *see* inequity and consequently address it.

Primacy of Homeownership and Built Environment

Another disaster response logic was the progressive steering of services toward the needs of homeowners and the simultaneous scaling back of services for low-income and nonworking economically deprived survivors. During disaster response, there is great emphasis on the built environment. Therefore, catering to the repair needs of homeowners is paramount to the operations of disaster centers. This was evident in my interview with Caroline, site manager of an NYS-run disaster response center on The Rockaways, who stated:

> Rapid Repairs [the State-subsidized home repair program] was one of our highest service areas—you know, one of the service areas with the highest numbers because, you know, a lot of residential owners experienced damage. Right now, that is the big piece that's going around this agency, and this operation is making sure that landlords can connect to contractors and get Rapid Repairs to deal with some of the issues like boilers and, you know, mold and basement damage. That's a major thing.

Logic of Response versus Services

When I asked if data collected at the NYS-run disaster response centers was used to target the needs of residents, Caroline responded with a similar logic as did Beverly:

> There's no targeting at this point, um, not through this operation. This operation is, we are here, and we are manned, and we try to have every service that we think would support someone who needs disaster assistance. There's city, state, federal, and community organizations that are here, and, um, our central office is in contact with us daily to find out what kinds of trends we're seeing, what types of services are we seeing. There's been a change in the, uh, the types of services that were placed here, because there was no real need for them, you know, beforehand. Well, for a point, NYCHA [New York City Housing Authority, which oversees public housing]. They were here just for a short while, but no one was really utilizing their services.

When asked how they determined the needs of a community, Caroline responded:

> The need, you know, is based on the numbers because we're growing numbers every day and based on the activity with them. So, they know we're here. You know, so there are agencies that we didn't really need here. And so as, as we report on numbers, the central office would make decisions on that service is not really needed. And if there was a service that was needed, then they would bring that agency in.

Here Caroline does mention that she reports to her superiors, who scale back or add services based on her assessments of need, which she and they base on "numbers." This bottom-up approach to collecting data and providing services seems like a good practice, although it was not clear to me that low "demand" for services that cater to the economically deprived was necessarily a reflection of a lack of need in the community. Once again, the unstated, but problematic underlying assumption was that everyone had access to information about the services offered and was equally able to make it to these centers in a "timely" manner.

If we return to another statement by Beverly, there was also a race and class habitus mismatch regarding basement apartments, a common feature of New York City living, particularly among the working class and new immigrants. Beverly explains her confusion about these living spaces and the people who choose to make these their home:

> The sad part is, here, that most people are living in basements, and basements are not what the government ever considers what a person should be living in. I mean, I don't know because I'm not from New York, but it seems that everybody

rents a basement out, so many of them are illegal to rent out. Um, I mean, I don't know why people live in basements. It's kind of dark down there but they do, and they are very content. They have huge apartments in these basements. I know I always thought these were like a small basement. There are big homes down there. So, it's very different. It's very different. Each disaster in each state is so different.

Because I had been a working-class immigrant New Yorker who lived in the boroughs, basement apartments are not weird to me at all, but hearing Beverly talking about them in this way felt uncomfortable. I lived in a New York basement apartment with my mom and sister for years, so her statements made me hypervigilant of the difference between our class status. While basement apartments are not ideal, they are common and usually the most affordable unsubsidized housing option that one could find in New York City.

Beverly's statements did reflect sympathy for these disaster survivors who resided in basements, but they also reflect the cultural bias stemming from living within a White middle-class habitus, an environment imbued with White middle-class cultural tastes (Mayorga 2014). This was an unfamiliar environment for her. She told me that she is from a midwestern state in a middle-class neighborhood with single-family homes and where residents don't live in basements. It is therefore hard to relate to residents who live in basements. This idea of a mismatch in habitus was also relevant to the issue of nonlocal adjusters of different racial demographics, devaluing the losses to property in basements of Black immigrant disaster survivors as discussed in chapter 4.

Invisibility of the Self-Employed

Yet another example of a disaster response logic was toward formal employment. Self-employed disaster survivors did not receive compensation for their trade tools, such as a DJ's CDs and records or a mechanic's tools, because the application did not provide a means to categorize these correctly. Similar to discussions in chapter 4 about the inspection of adjusters subcontracted by FEMA, a common complaint from low-income basement disaster survivors was that their belongings did not receive the correct valuation. Many felt that adjusters were of a different race and different socioeconomic status, and that these differences colored their subjective judgments about the value of the survivors' belongings and what they ought to be able to do without. This mismatch in race and class of responders and residents led to distrust of subjectivities that potentially obscure the assessments of survivors' circumstances.

Naquita, whom we met in chapter 4, explains that FEMA's IHP program, although it covered loss of personal items, did not cover her husband's losses associated with his self-employed status. In this instance, Naquita explains that her husband is a DJ:

NAQUITA: All of his thousands and thousands of music [are] completely gone.
SM: Did you put that in your application?
NAQUITA: Yes. And FEMA is like that's not their responsibility. It's just to get you started again.
SM: But did you mention in your application that he was self-employed?
NAQUITA: Yes, yes.
SM: Okay.
SM: And nothing?
NAQUITA: Well, a colleague of mine has a family member that works for FEMA, and I spoke to her on the phone, and she also said, 'You need to make sure when you appeal, they know that he is self-employed and that's how he made his living.' And, you know, I spoke to the FEMA representative here—
SM: They said the opposite, right?
NAQUITA: Yeah. He was just like, 'Oh, you need to take that all out.' He actually crossed it off of my letter.
SM: Yeah. [*nods*]
NAQUITA: So now we just wait and see what happens.
SM: One more thing, how long after the storm did you put in your application?
NAQUITA: The next day. It happened—the Monday, I think it was. Tuesday morning, when we came home, when we saw what was happening with neighbors, we actually put it on that same day.

Naquita's final statements reflect that despite beginning her application process early, holding a master's degree in organizational psychology, and possessing social capital that affords her informally transmitted information through her friend's relative who works with FEMA, she was still caught in the labyrinth of appeal, discussed in chapter 4 and via her husband, the exclusionary logic of response along with the latecomers who are not as endowed with capital as she is.

Ricky's Story

What happens to a fifty-year-old Black man who is chronically economically deprived and precariously connected to State services through a drug rehabilitation program after disaster strikes? What does it mean to his life that an *ecology of in-*

equity had ushered in a logic of response that displaces the logic of services? Ricky had lived in a three-quarter house in a pocket of deep poverty in Eastville, where, as Ricky describes, when food from the grocery store washed out onto the streets with the fast-moving toxic floodwaters, "people were actually picking those hot dogs up and eating them." Although my interview with Ricky took place in Eastville only a few days after Sandy floodwaters displaced him from his dwelling, follow-up interviews with him took place at the Resiliency Is Us disaster response center, where I encountered him again.

There is a bit of irony in Ricky's story, in that even before Sandy, Ricky was already living in transitional housing. FEMA offers transitional housing to disaster survivors who had experienced displacement. He became homeless and spent much of his time going to the Resiliency Is Us Westville disaster response center to keep warm and have a meal, keeping his appointments with social services counselors and caseworkers and interfacing with FEMA. The part of Ricky's story I was able to capture ends with him back in transitional housing.

Ricky's experience illustrates how chronic crises become enmeshed with acute crises during a disaster event and how taxing it is on the most economically deprived survivors. Although Ricky is in contact with FEMA immediately after the storm, his social location presents several obstacles to receiving resources in a timely manner. A governmental program would later transfer Ricky to transitional housing, exactly where he began. He was swept into a cycle he had almost prophetically described to me earlier in great detail.

Through his own words Ricky talks about the cycle of poverty and his frustration and lack of confidence in "the system" that he sees as not really designed to help him find a permanent solution to his housing precarity, not just for him but for countless others like him. His story also shows how ill-equipped disaster response is in dealing with persons caught in cyclical socioeconomic resource deprivation and displacements. The first time I met Ricky was on a street corner in Eastville, as he was helping load a soggy mattress onto a truck from a damaged halfway house tagged as condemned after Sandy's storm surges had engulfed it. I asked Ricky how he was affected by the storm, and he indicated that before Sandy, he lived in transitional housing as part of a drug rehabilitation program. Like everyone else, he incurred losses due to Sandy. However, experiencing Sandy would be more consequential for him than those who are not Black, male, chronically economically deprived, and substance dependent.

Sandy has made him homeless due to his housing precarity even before the storm. Although he started the application process early, he is having a tough time getting into a FEMA hotel, although his White girlfriend, who is similar to him in all other respects except race and gender, is staying in one. Because he can't stay there with her, she comes out to keep him company, and so they walk the cold streets together. Ricky and I discuss his situation. Ricky tells me:

> RICKY: That's a three-quarter house. You know, water damage, everything, got done to it. So, I lost my clothes. I lost stereos, everything. So yeah, I'm really affected by it. And I'm still looking for, you know, [housing] placement. Because every time, you know, I try to go, you know, to the FEMA hotel, they don't have any. And I'm, and I'm really stressed out about this, you know. And, um, I'm walking the streets, and I don't like to do that. That's not me. That's not me.
>
> SM: And you lived in that house.
>
> RICKY: That's where I used to live at.
>
> SM: Okay, describe what type of housing it is.
>
> RICKY: It's a three-quarter house, I guess, for like people that have drug problems, you know, trying to get their life together. And that—that's what I was doing, you know. I'm still doing it. So, I'm doing outpatient for all that. I'm not going to give this up. Because, you know—actually, I'm tired.
>
> SM: So, when you say you tried to get help, when you go, do you call them?
>
> RICKY: Yeah, we call. I call and there's no hotels open, right now. And, you know, some of the shelters are full. So, me and my girlfriend, we walks the street, you know? I don't like doing that. I mean, I feel bad. I'm the man, and I can't—I don't like to see her out here with me. And that really, that really stresses me out.
>
> SM: They're saying that all the shelters are full?
>
> RICKY: When I call, they say they're full and whatnot. So, you know, I'll just deal with it right now.
>
> SM: They told me that FEMA actually was out here. Did you actually see anybody?
>
> RICKY: Yeah, yeah.
>
> SM: From FEMA?

Ricky had his first encounter with FEMA almost immediately after the storm because they were physically present near where he lived at the time, but now they relocated somewhere else.

> RICKY: They—yeah, I seen—I did the application. They came to the house. They seen everything that was damaged and whatnot—to my room and all that. And I qualified for just about everything.
>
> SM: Okay, good.
>
> RICKY: So right now, I'm waiting on them.

SM: And how long is the wait? Did they tell you?

RICKY: Uh, probably about seven to ten days—about five to ten days for the money or whatever—replacement money. It ain't gonna be much, but it's something to get me on my feet. You know?

SM: Right. And then, during that time, they didn't say what else you can do?

RICKY: Eh, no.

SM: It's just waiting?

RICKY: Yeah.

SM: And how long did it take for FEMA to actually come out here?

RICKY: It took them like—it took them—they came out quick, I mean like three days, four days. Yeah, three days. It took them—they came.

SM: But it's just the process of having to wait for them?

RICKY: So, in the meantime, you see what I'm doing. I'm helping. Helping to keep the money in my pocket, you know? Other than that, I'm good.

SM: Do you know any other residents who were displaced as well, and they're looking for a place to—

RICKY: Well, my girlfriend, she's on the next block. She's a White woman. You can talk to her. Tell her Ricky sent you. Her, her, her—I call her my wife. [*Smiles*] Just tell her that her husband sent you. [*Smiles*]

Two months later I met Ricky at the Resiliency Is Us disaster response center. At this point, he is still homeless. He explains the issues he has been experiencing with his FEMA application despite starting the process so early. Many of the interruptions relate to his chronic conditions of poverty and his recent homelessness. First, he has issues accessing mail. The FEMA application process assumes that an applicant has a mailing address. I asked Ricky what happened since his FEMA application and reminded him that he was waiting to hear back since we last spoke. He tells me:

> FEMA did give me a runaround. They haven't sent me my check. I mean they sent it to the address, but it went back. So, what I gave was—I changed the address to a drop program with my outpatient program. And they said they sent it and it was supposed to be there. I went there today to find out from accounts. Say I need an additional hold on the thing. He said, "If you can do it like that, it's only one time, because it's like an emergency." They allowed my mail to come in.

Ricky continues to explain how both acute and chronic crises dovetail in his life. Even after two months, little about his circumstances have changed. He is still homeless and hopes to find permanent housing for him and his girlfriend, who he shares is now expecting a baby. He describes his feelings of powerlessness hav-

Logic of Response versus Services

ing her on the streets with him and not being able to secure housing for them and their baby. He explains the psychological toll his housing situation is taking on him. He said he was experiencing frequent triggers that lead to violent outbursts of which he is not proud. Ricky attributes this to his disaster experiences, which left him traumatized. He and his girlfriend had witnessed the wall of water that came down the streets of The Rockaways. He tells me:

> RICKY: Right now, I have nowhere to live. Okay, you know. I'm not ashamed to say it, you know, 'cause I'm still waiting on this money. This way, I can get me an apartment, right along with my fiancée. This way, we won't have to be out here in this cold. I'm not proud of what I'm doing, you know. She's out here with me, and, and that makes me feel small, and she's pregnant.
>
> SM: Oh no.
>
> RICKY: That—that makes me feel like this [*holds thumb and index finger together indicating 'small'*], you know. I don't like that. I'm still going through the bad experience, you know. I've had some time—sometimes I have attitudes, you know. She says something to me, I snap. Maybe I don't mean to.
>
> SM: Is that happening more after the storm?
>
> RICKY: Yeah, more after the storm. It's just—I don't know how to deal with this. I ain't never went through these storms before. Where I'm from in New Jersey, we don't have this type of stuff.
>
> SM: Right.
>
> RICKY: This was a shock and a surprise to me. I didn't even think it was really going to happen like it did, but it happened.
>
> SM: All right. Did you get to see the storm happen?
>
> RICKY: I was out—I was out in the storm. Me, me, and my fiancée was out in the storm.
>
> RICKY: With the water?
>
> SM: Yeah, watched the water come right past us, you know, and—
>
> SM: Do you remember that a lot?
>
> RICKY: It's still on my mind, yeah. Next day, you wake up, you find everything destroyed. Tables floating down the street, uh, people's homes destroyed. Hot dogs floating down the street.

Using his FEMA voucher becomes a complicated process because the FEMA process makes several underlying assumptions about applicants that do not apply to people experiencing the level of poverty and marginalization that Ricky has. One assumption is that applicants have access to a credit card and a state driver's license or identification card. Many economically deprived African Americans do not own these forms of identification. Owning a credit card is certainly out of reach for many of Eastville's economically deprived, including Ricky. Credit is also a function of race. Ricky mentions that his girlfriend has a credit card. Despite her being in many similar circumstances as he, owning a credit card and being able to stay at a hotel are two things she has access to that he does not. I ask him about his current housing situation:

> SM: So, you told me you were not sure where you were going to be able to stay after. Did you find a place to stay?
> RICKY: Well, FEMA they do the hotel. Okay, they set you up with a hotel. Any FEMA hotel. They've got a deal with FEMA. They can tell you they're gonna pay for the room and taxes. But what they don't tell you that when you get to the hotel, you have to pay a deposit out of your pocket. They don't tell you this. So, you get there without the deposit, they not going to let you stay in the hotel, regardless if FEMA is paying the money—you know to stay in the room. And I, just in case you wanted to use the bar, or the snack bar, or the phone, or something was wrong with the room, they [can] keep that money. But, you know, you use nothing, they give you that money back. Being that I don't have a credit card like she does [*points to his girlfriend*], or I have a picture ID, but my picture ID is coming from my benefits, like from HRA [Human Resources Administration].
> SM: You don't have a driver's license or a State ID
> RICKY: No, none of that. No, I don't have none of that.

Yet, another underlying assumption of FEMA is that an applicant has access to communication. Keeping minutes on his phone was an issue for Ricky. He expressed that he wished FEMA had maintained their physical location in his community. This would have facilitated a smoother follow-up process and would have reduced his bureaucratic burden significantly. I ask Ricky:

> SM: And you still haven't heard from FEMA yet?
> RICKY: No, I haven't heard from them yet, but eventually it's going to come.
> SM: And you're calling them via phone?
> RICKY: No. Right now I have a free phone, but I haven't got any more minutes. So, when I do get this, I'm planning on getting an unlimited phone, so I don't have to worry about running down the minutes. If FEMA was here [in a physical location] now, I would go talk to them, see what's going on. I'm good right now. [*self-motivating talk*]
> SM: Okay.
> RICKY: I'm dealing with it. [*self-motivating talk*]

Logic of Response versus Services

Ricky describes the point at which disaster response and social services collide, transitional housing. He is accustomed to having to navigate short-term deadlines and the precarity of knowing that he may be back on the streets at any time. He makes it clear that all FEMA will do is give him extensions and a check. His own longtime quest for permanent housing, which is not a mission of FEMA, is what determines what he does with this money. His vision is to use it as a stepping stone into a permanent place to live.

> SM: So right now, you're still at the FEMA hotel. How long can you stay there?
> RICKY: Huh?
> SM: How long can you stay at the hotel?
> RICKY: Maybe through January 13th or 14th.
> SM: And by then, they'll give you something more permanent?
> RICKY: They'll give me an extension. They'll keep giving me extensions until I—well when I get the money [*inaudible*], it's called transitional housing—I'm still going to be able to take that money and go find an apartment. That's what I plan on doing. I'm not planning on going and blowing it on nothing else. I've got to get out of the cold, you know. I can't stand this cold. Like I said, after, you know, after, the alcoholism departments will, you know, drug, alcoholism, supposed to help you get your, help you with an apartment, you know, find housing. But sometimes they do, and they don't.

Ricky mentions that after thirty days, one graduates from these drug rehabilitation programs, in which he is enrolled, so I ask him what typically happens after thirty days. He expresses his disbelief that after these men who enroll in these programs get to their graduation milestone, they end up back in shelters. He explains that spell of homelessness occurs because the city had not yet found them permanent housing. He indicates that some opt to go back to the streets and start using drugs again. He stated that they do so because they become despondent when they learn they will not receive permanent housing, especially after working so hard toward that goal:

> RICKY: Um, after the thirty days, if you don't have nowhere to go, they'll refer you to a shelter like Bellevue Shelter—which I think is really crazy, because the fact is these counselors down here supposed to be helping with this. We're supposed to be able to have an apartment by then. HRA will pay for this until you get on your feet, but they just do what they want to do sometimes. Like VNS. Visiting Nursing Service. Now they're real good people. They stand for who they say they are. They don't leave you out in the cold. I had them at one time.

SM: What happens with a lot of people who don't get permanent housing after the thirty days?
RICKY: They hit the streets.
SM: And then you think they go back to—?
RICKY: They go back to what [they] was doing—using again.
SM: Okay.
RICKY: Because they can't deal with it.

Ricky illuminates that this theme about finding permanent housing is a never-ending quest for economically deprived men who enroll in these drug rehabilitation programs. He skillfully explains the relationship between unfulfilled promises to the economically deprived and the cycle of homelessness, drug addiction, and relapse. Ricky launches into full advocacy for himself and others in his situation, who are yet to find permanent housing through their drug rehabilitation programs. He feels like these State programs continually make promises to the economically deprived, only to repeatedly fail them. He sees himself and others in this system, as treated as mere cogs in this machine:

> But you [would] think they was getting help from these people [housing services staff]. These people let them down, and they ain't got nothing else to worry about. [So] they say, "Bump it. Hey, I'm going back to doing what I'm doing" just to repeat the cycle all over again. You go back out here, and you go back to detox. Go back to these houses or they just keep on moving saying, "It's no use going back to these places." I can't—I've never been to that point, but you know if you're going to help, if you say you're going to do, be real for what you're going to do. Okay? You going—you're saying you're going to help these people, you say, "Look, I'll try to get you, you know, permanent housing by trying to graduate," make sure that you help these people get these permanent houses. You know? Don't wait till they graduate, then after they come out [of] the house, they got thirty days to leave the house, and then they got nowhere to go but to go on the streets and then they go back to using again. So that's why I sometimes I believe in the system, and sometimes I don't. That's how I am. It's out of my control.

Ricky then explains the elusive process of finding permanent housing. He explains that because the $2,700 he will get from the FEMA grant will only cover the first month, it doesn't quite get him to permanent housing, his end goal:

SM: Do you think this storm is going to help you speed up the process with helping you get permanent housing?
RICKY: Um, I can't say. Maybe, maybe not.
RICKY: If I have to do the footwork on my own, I ain't got no problem with it

but the money, if my FEMA comes, I'm banking then see what I can find underneath that—in that new—that range, $2,700. I mean, they expect me to find a property with that type of money and expect to keep it? Come on, that's one month's rent. That's one month's rent! What are you supposed to do after that, go back to the streets? I don't know, but in the meantime I'll find something, even if it's a basement apartment to start out. I don't care. We'll find something even if it's a basement apartment, I don't mind. I'll start out small and work my way up. Yeah it's only $2,700.

SM: And you have to figure out how to pay the rest of it?

RICKY: Right, right. Even FEMA says well, if you send in the rent receipt to them, they'll pay it, but I don't believe that. You're a fool. You ain't going to make me believe that. Come on. You can pay my rent until I get on my feet? Come on, uh-uh it's hard for me to believe that.

When I last contacted Ricky, he still had not found permanent housing. He stated that a government program had transferred him, along with several others, to a motel in Brooklyn. He had not found his version of the American dream, which was to no longer be homeless. He wanted to be off the streets and out of the cold. He wanted to be free of the cycle of drug rehabilitation programs that traps you. He no longer wanted to be in transitional housing but instead wanted a permanent, modest apartment where he and his girlfriend could welcome their baby and be off the streets. He also still had the dream of upward mobility, thinking he could work his way up out of poverty.

I later lost contact with Ricky, as he did say his phone was one of those pay-as-you-go phones. I don't know if any of the aspirations he had for himself ever materialized. I don't know whether the *logic of services* or *logic of response* rescued him from chronic poverty, substance dependence, and housing insecurity. From our past conversations, he believed that both were failing him in important respects. If he never makes it, it won't be for a lack of trying. I do know he had extraordinarily little confidence in the system that has let him down and so many others like him, whether it be the services arm or the disaster response arm of the state. It is unclear at what point his persistence ceases, if ever.

CHAPTER 7

Social Capital Privilege

> The key to this is the center. That's where people get their information. Information is the key to it. It's where you found out what you needed to know, where you got the latest information from.
>
> —Joe, Westviller, Westville Resiliency Is Us Center

In this chapter, I draw on my observations of the participation experiences of Westvillers and Eastvillers in a momentarily "desegregated" interactional space of the Westville disaster response area in search of disaster aid. I also contrast this site with other sites in Brooklyn, which provide counterfactuals to leverage my conclusions. The two categories of Rockaway residents, which I have given the pseudonyms Eastvillers and Westvillers, occupy long-standing, divergent economic realities of residential clusters bearing distinct racial and socioeconomic demographics.

Resiliency Is Us Incubates Bridging Social Capital

The "thirty-thousand-foot view" of disaster response activity at the Westville Resiliency Is Us disaster response center and surrounding area may have resembled an ant colony carrying out an egalitarian distribution of disaster resources. A casual observer might have concluded that disaster response organizations create networking spaces that buffer against the forces sustaining the enduring material inequality of the external environment. The Westville Resiliency Is Us center arranged in plain sight food products, winter clothing, cleaning supplies, and printed information about government aid programs. I did not discern major coordination issues at this center, beyond the late arrival of Resiliency Is Us to The Rockaways. Friendly volunteers were eager to provide relief and recovery assistance to whoever made it to the Westville disaster response center. My field notes of resident perceptions offered no overarching suspicion or particular concern with NGO discrimination or corruption. Initial observations and face-to-face conversations with Westville NGO "patrons" would support the notion of the egalitarian idealism one might expect of disaster response centers.

I waded through the sand-dusted streets of Westville and happened upon this make-shift center toward the end of a well-traveled street in Westville. Over the course of my fieldwork on the peninsula, I would repeatedly return to this fulcrum of disaster response activity. In the early wake of Sandy's landfall, I peered into the distant gazes of shivering Rockaway disaster survivors who poured out of their cold and damp homes, making their way past piles of rubble, and finally were received into the warmth of the generator-powered, Westville disaster response center. Disaster survivors dragged their tired bodies along to serving tables, where they would eventually clasp their frigid fingers around thin cups of hot beverages and plates of hot, cooked meals. Once fed, thawed, and swaddled in ill-fitting winter coats, disaster survivors would delve into stacks and rolls of relief supplies, often at the beckoning of the center's field manager or an eager NGO volunteer.

Joe was now nursing the cup of coffee that a Resiliency Is Us volunteer had poured from behind the row of serving tables lining the far wall of this makeshift NGO disaster response center nestled in this affluent, homeowner community of Westville. Just a few days earlier, the Atlantic Ocean had breached its shores and engulfed the basements and first floors of homes and buildings that lined this and surrounding Westville streets. Joe explained, "The key to this is the center. That's where people get their information. Information is the key to it. It's where you found out what you needed to know, where you got the latest information from." The structure abutted the steel fence enclosing the repurposed court on this well-traveled street in Westville. Resiliency Is Us had launched its disaster response efforts on the peninsula from this makeshift site. This site had morphed into this lucrative fulcrum of disaster response activity for Westvillers.

Joe's statements hinted at the significant role the Westville Resiliency Is Us disaster response center played in the early phase of the post-disaster recovery of Westvillers. Joe was a Westville homeowner and landlord of rental apartments in Eastville. Inside this makeshift NGO disaster response center, Joe and I sat at one of the long white event tables and chairs, fit for ten guests. I hunched over across the table toward Joe, as I listened intently to firsthand accounts of his subjective experiences with disaster assistance and recovery. I marveled at the calm and certainty of recovery he exuded, as he relayed in past tense the chaotic, uncertain beginnings of what were now successively accomplished disaster recovery milestones, which he attributes to NGO-mediated disaster assistance. Joe's disaster narrative stood in stark contrast to those of Eastvillers I would meet later that day. Many Eastvillers I had met, spoken with, and observed by that time had not yet experienced such successive bursts of progress toward recovery, nor would they ever experience such, even several months into the official disaster response on The Rockaways.

Joe's emphasis on information illustrates an interpretive use of the Westville Resiliency Is Us disaster response center as a repurposed site of interaction for accessing pertinent disaster information, not mere material supplies. Joe's own experience of gaining access to "the 'latest' information" confirms the centrality of receiving this information in his recovery trajectory. However, his awareness was not simply his entrepreneurial ingenuity. Joe's insight is rooted by his particular capacity to secure this unique access to a timely, constantly updating information stream that helped him navigate the shifting context of institutional support. In this respect, Joe's experience closely resembles that of other similarly situated Rockaway residents: these are the economically privileged White (in the context of The Rockaways, Irish) Westvillers.

Observations and conversations in the Westville disaster area confirmed that the key to gaining effective disaster assistance was the unencumbered access to nonofficial, informally transferred, time-sensitive information about disaster relief and recovery programs that often required immediate action. What produced this informal information stream? My observations of activities around the Westville disaster response area point to the emergence of a medium of information exchange: the informal relationships forged between Westvillers and Resiliency Is Us responders over a brief period of time. The clustering of Westvillers in conversing about daily developments and sharing information updates around the Resiliency Is Us center further created a feedback loop where information circulated among these conversing Westvillers in and around the Resiliency Is Us disaster response center, helping to incubate bridging social capital.

Through this word-of-mouth augmentation process, Westvillers benefited from early bird advantages, such as getting on a waiting list for repairs. This was significant because these repairs were worth thousands of dollars in savings from avoidance of assuming additional debt and preventing loss of equity in personal resources, which were common in areas outside Westville. A less tangible but real benefit was the lessened time, energy, and frustration spent on trying to navigate disaster-related programs. Westvillers were also able to influence some NGO decisions on the ground, such as keeping the Westville NGO center running, by communicating to responders the center's continued importance to the community.

Westvillers and NGO Responders Forge New Social Ties

According to social capital theory, a common type of *bonding* social capital stems from the familial or friendship tie, characterized by strong affect and expressiveness. The bonds created between Westville residents and volunteers exuded these qualities. In many ways the volunteers fulfilled a surrogate role of familial or friendship ties. In the absence of bloodlines and long years of routine relationship building, we can gauge the achievement of closeness and strength of resident-responder bonds, through expectations and displays of volunteers' emotional commitment and the psychological and material value to individual residents. We expect that familial and friendship ties are dependable and characteristic of an elevated level of commitment and certainly are self-sacrificing during periods of crisis. However, these ties form quickly and last for the duration of volunteer deployment ranging from a couple to several weeks. One indicator of the surrogating of this relationship is the trade-off decision to voluntarily miss one's own sentimental family traditions or one's own economic pursuits beyond the initial "sympathy" commitment to become a volunteer leading up to deployment to a disaster area.

The context of acute crisis catalyzes the process of forming these *bonding* social ties. The gravity of the loss and harm of surviving disaster sets much of the tenor of resident-responder relations. Several days after the flood waters had receded, Sandy's enormity continued to inscribe the facial expressions, intonations, sighs, long silences, and tearful accounts that communicated fragile emotional states of uncertainty, disappointment, hopelessness, and utter disbelief of misfortune. Volunteers from out of town responded to and even reciprocated these sentiments, sometimes moved to tears as they recounted personal stories of residents they spoke with, assisted, encouraged, and consoled. It is important to note, however, that this positive affect often responded to not just the needs of these disaster survivors but also their expectations. Responders' growing feelings of obligation to fulfill resident expectations, such as playing the role of confidant or conversation partner, all point to the beginnings of forging stronger, closer bonds in a shorter timeframe than is typical in routine environments.

In these respects, the particular context of the Westville Resiliency Is Us disaster response center conjured a quality of survivor-to-responder relations that presented as strong as, at least during the term of volunteer deployment, surrogate friendship or familial ties. At the person-person-block level, my observations and conversations with volunteers and Westville residents in and around the Resiliency Is Us center revealed that relations between Westville residents and Westville NGO volunteers were becoming less impersonal and transactional and more informal and personal. Verbal and nonverbal expressive aspects of interactions corroborated strong affect and empathy relayed in the mutual sharing of personal stories and even personal items. In the following accounts, I illustrate, through extended examples, how such rapid descent into a deepened personal connection occurs. First is an account of a forty-year-old, out-of-town Westville NGO volunteer with a background in emergency services. Here, she recounts her experiences with a Westville disaster survivor she met earlier that Christmas Day outside the Resiliency Is Us disaster response center, where we were now standing and conversing:

> This has been a blessing, and I can't expand on that enough especially today and seeing the people that are coming in here today and listening to their stories—uh—I was talking to a gentleman earlier. And after talking for a little bit, he said, "Can I come back and have a cup of coffee with you?" And I said, "Absolutely." So, he took his dog home and came back, and we talked a little bit more... And he spoke about his parents a little bit [also details about their background and his siblings] and then we talked about the devastation... but he also told me he wrote a book... and he was going to bring one back for me today and I hope I do see him. [*Pauses, smiles*] Very, very dear soul.

As the above account of the Westville NGO responder illustrates, the affect conveyed in the narration of such stories of the residents goes beyond the sympathy one feels when one sees a person in need. The thematic content of these accounts falls outside the scope of physical or psychological needs and solicitations or offers of disaster assistance. Instead, over the course of just a couple of interactions, the needle quickly moves from a transactional role interfacing to informal, interpersonal relating through successive disclosures of personal aspects of one's life. The NGO volunteer also personalized the interaction beyond interfacing through prescribed roles by identifying the survivor by name.

The NGO volunteer restates his request, "Can I come back and have a cup of coffee with you?," which denotes that his return was not simply to have a cup of coffee, but that he would be honoring the verbal commitment of returning to the tent to informally relate further with the particular volunteer. She responds "Absolutely," revealing that she had not found the request strange or intrusive, suggesting that the stories he had begun to share had moved the interaction toward more personal, less formal expectations. Again, during the second conversation, he offers her a gift of a book: "He was going to bring one back for me today." She also expresses in her account that she wants him to bring the book, and she says with a smile, "I hope I do see him," indicating an expectation that he will fulfill his promise to her. Not only is there an expectation that he will give her the book, but there is also the additional expectation of seeing him again. Her use of the term of endearment "very, very dear soul" suggests the interaction has achieved a degree of empathy and positive affect for the resident. This statement also reveals the NGO volunteer's confidence, or at least a mild level of trust, in the resident's character. That is, there is at least enough trust to allow her to attest to me, in his absence and without my asking, that he is a "dear soul."

Beyond the substance and tenor of relating, these NGO volunteer-resident relations required larger investments of time and physical and emotional energy, sometimes even at the expense of other familial, professional, or business commitments in the routine lives of NGO volunteers back in their home states. To continue with this illustrative case, the NGO volunteer immediately makes the following statement:

> It makes me so thankful for what I have. You know, today is Christmas Day, and I was supposed to go home today originally, and I asked to be extended for another week, so I'm going home on New Year's Day. And my daughter [and I] ... had Christmas all planned out, and it was kind of her first experience. She has her apartment and her car and her first Christmas in her new apartment, and she was really excited. And I called her, and I said this is what's going to happen. [She opted to extend her volunteer stay in New York.] And I'll be back. She said, "Mom, I'm so bummed you're not going to be around for any of the holiday season," and I said, "I'm going to know you for the rest of your life. Some of these folks need help now." She thought about it for a while and then said, "Yeah, ok, alright. I get it." I talked to her a couple times since, and then I extended, so I called her and said, you know, I was a little hesitant. [*Directed to me*] Because, you know, she was a little disappointed the first time. And said, "I've been extended for a week." She said, "Okay." I said, "Really?" She said, "Yeah, you know what? You really want to do this. This is what you want to do, and I support you 100 percent, and if you do come that day, I'll make sure I take off work and spend the day with you, when you come back."

Social capital theorists also define social capital as the expectation of and fulfillment of obligations. Another opportunity to uncover the nature of the relationships that were budding in this Westville NGO center is the phenomenological analysis of observations and perspectives of residents when expectations of commitment went unfulfilled. Residents' statements conveyed expectations indicative of a sense of a heightened expectation of obligatory personal commitment. For example, one resident who looked quite despondent, pointing to a bag containing two neatly wrapped gifts, stated his disappointment that two volunteers for whom he had bought gifts were no longer here and didn't even say goodbye. Such reactions convey a degree of emotional attachment discernible through the level of despondence associated with sudden loss of a close tie such as a friend.

The following appraisal of volunteers' presence and work illustrates a benefit that is distinct from the institutional support of disaster assistance. Such visible and audible expressions went beyond the usual commendation of service satisfactorily rendered but resonated a near-umbilical dependence. One poignant example of this distinction appears in the following statements of a sixty-year-old Westville resident as he discusses the distinct types of agency-mediated and organization-mediated disaster assistance he received:

> The volunteers was probably the most important because—uh—you never really felt that the government was here—to help us—yeah—the army was there periodically—yeah—you never felt—we felt like we were on our own—yeah—FEMA was here to help with financial things, but there was no sense of security that the government was stepping in or even the city trying to remediate these issues.

Even in the absence of such declarative comparison, residents spoke of the emotional impact of their interactions with volunteers. A sixty-year-old retired elementary school teacher enunciates in her classroom voice, "The volunteers are beyond belief. They're so hardworking, and they couldn't be nicer. They make my life very pleasant!"

The above examples of NGO resident-volunteer relations are not one-sided but reciprocal. Westvillers filled an emotional void for out-of-town responders. The sudden change in geographic location, beyond the experience of jet lag, led to a sense of "freefall" for volunteers coming in from other states. NGO volunteers showed signs of spatial and temporal disorientation. Many of these volunteers reported not having a sense of the time of day or the number of days elapsed. Some repeatedly expressed embarrassment due to a lack of general awareness of their geographic location relative to the impacted neighborhood, as well as a lack of local knowledge of surrounding streets. Some NGO volunteers also expressed not feeling equipped with locally relevant knowledge and skills pertinent to the particular disaster, despite having experience with other disasters. For these reasons, I posit that the forging of social ties between out-of-town NGO volunteers and residents was a function of mutual dependence and mutual capacity for providing emotional, informational, and social support.

How Information Diffuses through Westville Disaster Area

Daily conversations with residents served as both informational and emotionally grounding experiences for volunteers. Newcomer volunteers also learned from residents about new developments in the Westville disaster area. I observed Westville residents share with newcomer volunteers the problems they encountered using cleaning supplies and protocol the NGO provided them in the form of printed flyers, brochures, and forms. As weeks went by, residents shared their improvisations to these prescribed methods and relative successes. Residents relayed the developments with gutting their basements and the disappearance and later reappearance of mold. Volunteers learned from residents and turned around and shared mold remediation concoctions and rituals residents had perfected through trial and error. Westville residents who had tried nonworking numbers also shared alternate contact information as well as effective strategies for avoiding long wait times and reaching contractors. They also reported changes and their success with particular programs.

These "testimonials" served as a bottom-up information stream of updated information from the disaster area around the Resiliency Is Us Center where volunteers were stationed. This informal transmission of pertinent information became the centerpiece of this center. In particular, the volunteers were unwittingly becoming invaluable knowledge brokers of an emerging informal disaster assistance networking process as they relayed this updated information to other residents. The specificity, relevance, and immediacy of this informally relayed disaster information stood in contrast to the instantly obsolete and nonspecific information and guidance on the crisp leaves of flyers and brochures stacked on the Resiliency Is Us disaster response center's information tables.

The "unofficial" disaster information was location- and time-specific and instructive on acquiring and manipulating the specific disaster supplies circulating in the local environment, knowledge of delays and mishaps, selecting and securing repair services, and replacing large equipment, work tools, and lost personal items. Most significant was experience-tested information about maneuvering administrative hurdles to secure program benefits in short order, such as getting on waiting lists for programs with short application windows. Together, these resulted in *early bird* monetary advantages over others who were not privy to such privileged informational access.

Although Eastvillers frequented the Resiliency Is Us disaster response center and surrounding areas, forming social capital with NGO responders was an interactional accomplishment unique to patrons who resided in Westville. How did

Westvillers accomplish this feat? Observations inside the Resiliency Is Us center suggest that Westvillers who were patrons had a unique capacity to transform transactional encounters into informal bonds. I argue that this capacity rested in their relative ease of incorporation into this uniquely beneficial, mutually reinforcing, ecological relationship rooted in the placement of the Resiliency Is Us disaster response center near the homes of Westvillers.

Westvillers Capture Public Goods, Eastvillers Crowded Out

In the Westville Resiliency Is Us disaster response center, there were clear distinctions in the way that Westvillers and Eastvillers related to others in the Center. As illustrated in the previous sections, this difference is significant because these newly forged social bonds connect Westvillers to informational resources pertinent to their prospects of disaster recovery. The formation of social capital–yielding bonds affords *timely and advantageous informational access* to institutionalized resources during the crucial initial period of disaster response. These social bonds that disaster survivors strike with NGO responders are particularly significant because longstanding networks and their social capital become compromised due to displacement and dispossession. Large nongovernmental organizations, in comparison with small local community-based organizations, have greater and more direct access to governmental disaster resources.

The subsequent section illustrates some mechanisms that point to why Westvillers and Eastvillers had different opportunities for forging social capital–yielding bonds with NGO responders. Westvillers, the economically privileged White residents of Westville, employed symbolic *inclusionary projects* and *exclusionary narratives*. I argue that together, these projects and narratives helped set the stage for the interactional environment for deciphering deservingness and undeservingness regarding disaster resources by establishing parameters of legitimate and illegitimate claims and around presumptions of belonging and trustworthiness. I illustrate this in the following vignettes, narratives, and projects that Westvillers employed, which enabled them to forge strong affective bonds with nonlocal, NGO volunteers and the field site manager.

Westvillers Become Regulars, Eastvillers Remain Visitors

One of the engagements of Westvillers was *the project of becoming regulars* in the Resiliency Is Us center. Becoming a regular meant coming to the center every day, even several times a day, engaging in small talk and laughter, and relaying personal disaster stories. This undertaking involved learning the names of volunteers and ensuring that volunteers knew theirs, as well as using conversational cues such as "See you tomorrow" as they were leaving. Westville residents had converted the space of the Westville center to a street corner café.

Regulars knew the "waiters" and "waitresses" by their first names, and disaster responders knew what their resident patrons were "ordering" today. Of course, there was no money exchanged, but it had the ambiance of an old diner full of patrons. At times when I sat to interview residents, one resident, usually male, would perform a hand gesture toward the food, insisting I get something to eat as if he were saying to put it on his tab as his guest. I had qualms about eating because although I was there during the daylight hours, I knew I had the option of riding the bus back over the bay and grounding myself back into my reality of routine life, where I could order at a real restaurant or even cook if I was not too exhausted. I knew this was a public space, but the behaviors of Westville disaster survivors and responders blurred the lines.

On the other hand, Eastvillers who were non-White and economically deprived engaged the space inside the center as guests. They did not engage as regulars as Westvillers did. The racially minoritized from Eastville came to the Resiliency Is Us center, but unlike Westvillers most remained outside, where there was an array of visibly stored disaster supplies. Westvillers outnumbered Eastvillers in the Westville center. The few racially minoritized Eastvillers who ventured into the center walked past the information table and approached the tables where volunteers served them plates of food. Eastvillers sat at the tables furthest away from the traffic near serving tables. They did not interact much with other residents, and their verbal interactions with responders did not exceed responses to questions about what sides they wanted on their plates and if they had already received a coat. Eastviller engagement in the Westville center was basically "pack and go," as they often left immediately upon eating. Even for the few who stayed for several hours, there was still little to no interaction with other residents and NGO volunteers.

However, Eastvillers were not cultural isolates. Their limited interactions were location induced. The way that Eastvillers perceived themselves in the Westville urban space and the way they interacted within the Westville Resiliency Is Us disaster response center reflected the spatialization of race and class, split between Westville and Eastville. Samoa, a sixty-four-year-old Eastviller and Native American woman, had moved from North Dakota several years previously and had a clandestine living arrangement with her boyfriend in a single-room occupancy (SRO) paid by social services. Sitting at the Westville Center where she came for lunch every other day, she joked about the Westville residents. Samoa said, "They are filthy rich, I heard. In order to live here you have to be filthy rich, but look at them. [*Smiles and jerks her chin upward and toward them*] You can't tell they're rich. They look like you and me." Samoa had been a resident of The Rockaways for some years and yet had never traversed this Westville area before Sandy, reflected in her statement "I heard." She fixed her eyes upward, constantly looking away. Samoa was associating race position with class location in this context. She assumed that since neither of us was White, neither of us had Westville money. By her use of the pronoun "they," she also distanced herself from the experience of the impact of the storm on Westvillers. She continued to joke wryly that disasters are a consequence of sin and that the "poor" Westvillers had never had the experience of need, something all too familiar to Eastvillers.

Ricky's case presented some elements of anomaly: a Black male Eastviller and his White fiancée, who didn't wish to speak with me. Ricky, on the other hand, welcomed our conversations. First, they were an interracial couple in a place that many residents described as being a White Irish area. Several residents confirmed that Irish Whites of this area enjoyed the highest status and class position. The couple were also not homeowners. In fact, now they were homeless. Ricky was a sixty-year-old man who had lived in a three-quarter house in Eastville, where all his personal belongings were flood damaged. He and his fiancée, who was pregnant at the time, would come to the Westville Resiliency Is Us disaster response center and stay there all day, huddled together at the table at the center of the room. They did not interact with others, but they did spend long hours there.

I spoke with one of the mental health professionals in Westville, a sixty-year-old White male with an academic position from out of state, who volunteered with a large NGO as a mental health consultant for residents. He had mentioned that part of what he did was to initiate conversation with residents in order to determine if they needed mental health services. I later also had an opportunity to observe his interactions after we spoke. While the NGO responder invited himself to sit next to various residents and talk to them informally, this responder did not approach Ricky, although he would have been the perfect candidate for such services since he had endured chronic stress due to poverty, substance dependence, recent storm-induced homelessness, and numerous bureaucratic hurdles in order to find permanent housing. Ricky later told me that he would just "snap" at people for no reason—something he thought occurred more since the storm. In an environment buzzing with "small talk," no one engaged Ricky in conversation.

The distinction of who belonged to Westville and who did not resurfaced in several conversations and experiences in the area. I would come to understand that Black bodies, mine included, were hyper-visible in the Resiliency Is Us center. Carl, a Westviller and Resiliency Is Us center patron, relayed the unprompted story of the Black pastor who came to the center. He described him as neatly and elegantly dressed. He said that all eyes were on this man. The pastor dropped off bags of clothes and left. According to Carl, everyone stopped and stared at him, but no one went up to talk to him. Carl also mentioned that some residents were not happy to see persons from other parts of The Rockaway coming to serve them, even as volunteers.

"Clean" versus "Dirty" Exclusionary Narratives

The notion that Eastvillers were out of place when they trekked to the Resiliency Is Us center and surrounding area was most notable in my conversations with Westvillers. Carl recounts a scene outside the Westville center, where he thought some individuals were receiving unentitled assistance:

> People who weren't involved were taking clothes—which they really shouldn't be doing—taking away from the people here who needed it. People were coming in all dressed nicely. They weren't in this. People who were in this didn't take showers for days—we know who they are. People driving up taking water. Taking food that didn't belong to them. That was allowed to happen.

I would later hear this resonant exclusionary narrative, similar to Carl's, from a passenger on a bus ride from Eastville heading toward Westville. Iman, a White working-class male in his sixties, was informally and occasionally employed as a mechanic. At first, I thought he was from Eastville, but once he started talking, I recognized the familiar narrative. He confirmed my suspicion that this narrative was an early form of social closure when he said he was from Westville and patronized Resiliency Is Us but had only gone to Eastville to retrieve mail. Sandy's destruction had diverted Westville mail delivery to the Eastville post office. I engaged Iman in conversation, asking:

> SM: How did you find out about the Westville center?
> IMAN: When I was looking for socks at a dispatching center between Eastville and Westville. They told me, 'We got centers thirteen blocks down in Eastville or sixteen blocks in Westville.' I walked nine blocks from my apartment to the Westville center and nine blocks back.
> SM: Did you interact with other residents there?
> IMAN: Yes—and a few cheats who came back from Brooklyn and Queens who parked their cars somewhere else. That's when they instituted the rule that you had to show ID that you were from Rockaway.

SM: How did you know where they came from? [*Pause*] How did you know who came from the neighborhood and who didn't?

IMAN: Basically, because most of the people from Rockaway were dirty—Come on, how do you take a shower? How do you clean yourself up very well with no heat, no hot water, and some people didn't even have running water? The Rockaway residents—and no offense because I was one of them—were *dirty* even several weeks after the storm. These people were perfectly clean. Like they had taken a bath that morning—looked clean like they had taken a bath that morning. Obviously, you're not from Rockaway. How do you get to take a bath when 90 percent of Rockaway don't have heat, don't have water?

SM: Anything else?

IMAN: So, they started instituting that—after that everyone who came to the center had to show something valid.

SM: Any other differences from the people that made them stand out? The people from outside?

IMAN: Some of them took carts, but the carts were clean too. Your cart is dirty! 'Cause you can't clean that all the time. You're clean and you have a clean cart?!

Interestingly, after getting more details of the timing, I realized that I had seen firsthand on an earlier trip some of the events Carl was describing. As I walked up the street leading to the Westville Resiliency Is Us disaster response center and the Westville Catholic Church, I immediately noticed a small crowd of patrons outside. I counted upward of fifty women and men, mostly women, standing in line outside the church near the Westville warming center, where Carl would be talking to me on a subsequent trip. After speaking to some of these Westville Resiliency Is Us disaster response center patrons, it was clear that most were from Eastville. Among those I talked with briefly were West Indians who conversed in English and, among them, Haitian immigrants who were only able to converse with me in creole as well as Mexican immigrants who spoke primarily Spanish (indicating recency of arrival in the United States) and native-born African Americans.

Some of the women told me that they had heard that the Westville Catholic Church would be handing out food and basic supplies there. Some had heard from their neighbors and others from coworkers. Some had borrowed cars or received rides since they had lost their cars to the flood waters. What was most remarkable was that they were only now hearing about this location although it had been open for several weeks by then. I learned from talking with them that these non-White working-class and first-generation immigrants were there seeking supplies because the stores had closed due to flooding, and that their perishable food had spoiled due to lack of electricity. Their late arrival was a function of the slower diffusion of information through flyers and their networks in Eastville and the further distance of the apartment buildings where they resided.

Recall that neither Carl nor Iman ever mentioned that "clean" and "dirty" were proxies for race. Their categorizations were about who they thought belonged and who did not, and, by extension, who could lay claim to disaster resources and who could not. These boundary-making and counterintuitive symbolic narratives around the "clean" and the "dirty" indicated preliminary stages of exclusionary practices of social closure and hoarding. Despite the lack of race-specific language, which may be an artifact of my own racial minoritization, the most crucial distinction was that economically deprived Black and Brown Eastville residents were the object of scrutiny.

Carl and Iman were drawing a symbolic distinction between those they considered to be Westville community residents and therefore deserving "disaster victims" who bore the visible markers of loss and devastation and those who did not. According to Iman's account, the response to these events led to requirements of showing identification or proof of address, suggesting that these boundary-making projects had real consequences for residents who would not be able to furnish these documents.

Kacie, who had begun volunteering in the Westville make-shift center from its inception, indirectly confirmed Iman's assertion as notable. She told me that "Monsignor," the priest who oversaw some of the volunteer activities at the center, warned volunteers, "You don't say, 'You don't live below this street, so you don't come here.'" Such institutionalized rules would negatively impact Rockaway's formerly incarcerated and undocumented immigrants, who do not have these documents. These categories of residents lived in Eastville.

CHAPTER 8

Organizational Networks of High and Low Capital

> It started—it literally started by the community for the community. And then little by little, the mayor's office sent somebody down, and then they worked with us. Then Resiliency Is Us came, and they worked with us. Everybody's been very good. The military came down, worked with us. All of them are helping us.
>
> —Monsignor Paul, Westville Catholic Church

In this chapter I draw on my observations of and further inquiry into the relationships among local churches, FEMA, and New York State disaster relief and recovery centers as my basis for comparison across the urban areas of Westville, Eastville, Canarsie, Brooklyn, and The Rockaways. My conversational interviews were particularly fruitful in helping me gain a better understanding of how organizations come together to work on disaster response, from a bottom-up perspective. The chapter details the chain and order of events that attracted various organizations to these impacted areas, which are consequential for the relational environment of disaster assistance as illuminated in the chapters of this book. I present four models of emergence of spatialized, organizational, and institutional networks (*organization agglomeration*, *organization isolation*, *organization hosting*, and *organization coalition*) and discuss them in light of the disaster scholarship.

How Disaster Response Organizations Make Ecologies

While in the field I became fascinated with this question of *response building* or how and why governmental and nongovernmental organizations and local community-based organizations come together to orchestrate disaster response in different urban areas. This question became even more significant when I observed differences in organizational presence within The Rockaways and between The Rockaways and Brooklyn. What was noteworthy was the lack of governmental and NGO presence in one of the most disaster-impacted areas in Eastville. One important aspect of resource inequality across disaster-impacted urban areas is inequality in the capacity of certain areas to attract and establish a central base for relevant governmental and nongovernmental resource distribution.

My fieldwork supports what we know from the disaster literature: media coverage is important to drawing attention to and consequently drawing resources to more economically privileged urban areas over economically deprived or darker ones. However, I uncovered another response-building mechanism that helps stream pertinent information and resources to some communities at the exclusion of others. This higher-order mechanism of inequality connects institutional, organizational, and spatial levels and is crucial to creating what I describe in this book as *ecologies of inequity*. An ecology of inequity emerges through on-the-ground response building among disaster response organizations, which induces an emergence of an *ecology of privilege* in one area, while simultaneously relegating an *ecology of disadvantage* in a neighboring urban response area.

In Westville, I identified what I am calling *organization agglomeration*, a chain reaction process by which a network of large and smaller disaster response organizations becomes concentrated around a large NGO nucleus in a disaster response area. I uncovered the opposite in Eastville, where Always With You, the local community-based organization, remains outside this resource network. Resiliency Is Us established only loose, fractious relations with Always With You, which relegates the latter to what I call *organization isolation*. I identify another type of response building in Canarsie, namely an *organization hosting* where the Catholic church merely provides the space to host the local government efforts. Finally, I identify more decentralized networking among local community-based organizations and large and small churches among numerous Brooklyn urban areas, which I call an *organization coalition*.

Westville: Organizational Agglomeration

It is no surprise that more affluent, usually majority White, middle- to upper-class urban areas have better institutional resources than urban areas of concentrated disadvantage, which are almost always predominantly racially minoritized urban areas. What is less clear is how and why these neighborhoods accrue more institutionalized resources after a crisis than economically deprived neighborhoods that may need such resources most. By comparing the organizational relations and presence in Westville to the dearth of organizational presence and collaborative relations in Eastville, it is evident that some urban areas are better draws for large, high-capital NGOs than others.

In the disaster response area of Westville, there were mobile trucks from a variety of organizations, volunteers, food trucks, and mobile health clinics all providing essential services and resources to Westville residents. This created a "nucleus of relief" and a "busy-ness" or "buzz" in that part of Westville. I was curious as to why there was this disparity across the corollary disaster-impacted area in Eastville, which was only a few blocks away. After interviewing key embedded actors of local churches, community-based organizations, and NGOs on the scene, I realized that this was not the result of a centrally coordinated effort, but the result of a chain reaction formation of organizational ties, a key aspect of *organization agglomeration*.

A casual observer might assume, regarding the emergence of disaster response, that a "lead" governmental organization would solicit the help of a number of organizations and deploy them to a particular location where the need was determined to be greatest or unmet. However, I quickly learned through my interviews with church leaders who were heading major disaster response operations in Westville and also Canarsie that the process was much more diffuse and even happenstance. I draw on my interview with Monsignor Paul, a sixty-year-old local priest of the large Catholic church and school spanning three blocks in the heart of Westville. Monsignor's church spearheaded an effort that ended up serving, as he told me, over "ten thousand residents per day." He explained:

> MONSIGNOR PAUL: I started maybe two days after, uh, the storm. Some people came. Young people with clothing. Uh, it started with a very small room and ended up with the whole school building, and then food, and then the government kind of attached itself to us little by little.
>
> SM: How did they contact you? The government.
>
> MONSIGNOR PAUL: They showed up.
>
> SM: Oh, literally?

MONSIGNOR PAUL: It started—it literally started by the community for the community. And then little by little, the mayor's office sent somebody down, and then they worked with us. Then Resiliency Is Us came, and they worked with us. Everybody's been very good. The military came down, worked with us. All of them are helping us.

SM: So, how did they get to hear about you?

MONSIGNOR PAUL: Uh, we were really the only ones on the Peninsula at the time doing anything. And so that's why we ended up with ten thousand people—ten thousand people a day.

SM: Okay. And so all of the organizations that were here? I did come here a few days after the storm, and there were lots of different organizations—

MONSIGNOR PAUL: Well, they came—around that area—they came little by little. Uh, it really started by the community for the community, and then little by little, things got added on. Uh, as we worked, uh, a structure got put up maybe a week later. Uh, and then everything was added—a heating tent, and then the food, and then—whatever else we needed, we just added on to it.

SM: Okay. So, it's not like you had preexisting relationships with organizations?

MONSIGNOR PAUL: No, no. They just—it just kind of happened. We never had anything like this before, so.

The Westville example illustrates how several governmental and nongovernmental entities ended up creating this "nucleus of relief" in Westville. Monsignor Paul describes a bottom-up process of how random acts of kindness mushroomed into a massive disaster response effort that ended up serving thousands a day. His "by the community for the community" emphasis is the classic bootstrap story that stresses the assertive actions of average White, economically privileged citizens that eventually lead to monumental achievements.

However, by comparing this process and outcome to Eastville, this is as much a story of the emergence of organizational ties and pooling of resources in one area versus another area with equal or greater need. As the monsignor pointed out, the local and federal governments had quickly supplemented Westville's *crisis capital* once community efforts were already underway. How were they so successful when I had seen and heard similar "initiating actions" by small immigrant churches in Brooklyn and Always With You in Eastville without such supplementation?

Monsignor Paul's account illustrates the creation of *social capital* at the organizational level, beginning with individuals within a large, well-resourced organization, in this example a Catholic church, which eventually attracted the mayor, a large NGO, and the military. The social capital literature talks extensively about the value of organizational ties in brokering the social capital of individuals (Small 2009b), but here we actually see how this occurs temporally and how embedded actors can help forge bonds across organizations. Monsignor Paul also described these emergent bonds as collaborative. This description of the collaborative and consistent presence of governmental and nongovernmental organizations by Monsignor Paul is a stark departure from the descriptions that low-capital small immigrant churches and community-based organizations in Eastville and in Brooklyn offered. Leaders of these low-capital local organizations described the disaster response experience with these entities as fragmented and episodic. Some low-capital community-based organizations even expressed less than amicable relations with large, high-capital NGOs when they sought resources from them for their communities.

My interview with Pastor Ward and Bishop Fabian, two church leaders of Caribbean background serving Canarsie and other Black immigrant communities in Brooklyn, reveals some of the challenges these local church responders and residents of majority Black areas faced. These challenges existed despite experiencing similar levels of flooding as more economically privileged White areas in New York City. These church leaders expressed dissatisfaction with the level, type, and timing of assistance the community received from high-capital organizations such as FEMA, NYS, and Resiliency Is Us. The ministers attributed the disparity in response to race as well as the fact that local officials had not done enough to sound the alarm that Canarsie had suffered a similar fate to some of the more publicized disaster areas. Bishop Fabian, who heads Brooklyn Black Church Consortium, discussed his disaster-relief work:

> Good afternoon. My name is Bishop Fabian. I'm the pastor of Never Lose Hope Fellowship. We are here in Canarsie providing food and clothing to the residents who have been displaced and suffered loss from the hurricane. Unfortunately, um, they have been forgotten. They have not gotten any resources. The elected officials are not here. FEMA is really not here. FEMA is only filling out applications. There are people who have no power, who have no lights, who have no water, and this community has been forgotten, so the church has responded, and we have a lot of volunteers. We have workers here all the way from North Carolina [referring to the disaster relief supplies my family organized] and still we haven't seen the elected officials from this area responding to this great movement.

Bishop Fabian pointed to a two-week lag in the response to this Black immigrant enclave. He lamented that FEMA, the city, and the State had not established a real presence in this community. In such a vacuum, these low-capital churches became

Organizational Networks

the lifeline for their communities. I spoke with Bishop Fabian about the delay in getting assistance:

> SM: What do you think is causing that disparity, because we just went to Staten Island, and we did see a lot of help get to some of the neighborhoods there.
> BISHOP FABIAN: I think several things happened. One is you have to have local leadership to rise up and make noise. If you don't make noise, you don't get the right response, and frankly, um, there are some racial disparities. You go to some neighborhoods where there are White folks, and they're getting it. You come to our neighborhood, and we're not getting the resources that we need.

Bishop Fabian understands the disaster response inequality across the neighborhoods he has surveyed and the Black communities he serves in as "racial disparity." He thinks the necessary response to such a disparity is strong, loud political advocacy. He places this burden at the feet of the locally elected officials. I spoke with Pastor Ward, who has partnered with Bishop Fabian on the disaster response efforts in Canarsie. I ask him:

> SM: Has any help gotten to the neighborhood?
> PASTOR WARD: Well, the help is scant in some areas and nonexistent in others. FEMA has been around, but what FEMA is doing is registering people, making sure people register with FEMA so they can get help, but what we are concerned about at West Indian Healing Church is people's immediate primary need. What people need now.
> SM: And what do people need now?
> PASTOR WARD: What we're finding is that their homes are flooded, they have no electricity, no heat, and no light. What they need is that: pumps and generators to pump the water out of their homes. They need restoration of power. You know, they need a hot meal. Some people have lost their belongings, so they need clothing. You know, so those are the things that we're trying to—we're trying to bring into the community, and we're so glad that you guys are here today to assist us in doing that because what you're doing is more than what the governmental organizations are doing. The mayor's office, for example. In some areas what they do is bring in military food in the military package and give them out to the people, but some people have gone for two weeks without a hot meal. However, in some areas you might see, you know that they're giving out hot meals. Resiliency Is Us is not as prevalent as we thought that Resiliency Is Us would have been, and also Saving Grace [a national relief organization].

Pastor Ward then stated that he had heard that these large nongovernmental organizations were prohibited from going to an economically deprived Black eastern part of The Rockaways.

> SM: Why do you think some neighborhoods are getting help and some are not?
>
> PASTOR WARD: Um, it all has to do with—well, according to, the official thing is, it's priority. The other thing is, depending on which neighborhoods, the more affluent neighborhoods are being helped before. You know, again, it's—it's a disparity that exists in our community, and the thing that we as Black people have to live with. You know, I guess we don't have to live with it [*chuckle*], but it's a fact of life, what we have to deal with.

Pastor Ward's analysis of the communities he serves is a bit more nuanced than Bishop Fabian's. He attributes the disparate response, the timing and consistency of response, to wealth disparities, yet for Pastor Ward these are not separate processes. He is acknowledging the racialized patterns of wealth disparity across majority White and majority Black urban areas. White urban areas, which tend to be wealthier, receive better assistance first, "a fact of life . . . we have to deal with."

Both Pastor Ward's and Bishop Fabian's accounts lie in stark contrast to the Westville organizational agglomeration model of official disaster response. Monsignor Paul and Pastor Ward both reference the mayor's office, Resiliency Is Us, and the military as being important actors in the official disaster response in their respective neighborhoods. However, the key distinction is the kind of neighborhood presence they establish. This contrast reveals the importance of the kind of presence large NGOs establish in neighborhoods. In some locations, Resiliency Is Us's only presence was through the visibility of their trucks as they came to drop off items to local organizations. Although media accounts and public outcry of disaster inequality across neighborhoods focus on absence or delayed deployment of organizations, it is not sufficient to assess whether organizations are present or disburse resources in communities, but whether they are stationed and *visible* in these communities as well as to what extent their efforts can be described as collaborative versus antagonistic or aloof with local organizations.

In the case of Westville, Monsignor Paul also pointed out that he did not seek out the collaboration his church received from FEMA and Resiliency Is Us, but that they "added themselves" to the church's ongoing efforts. How did these organizations know about the Westville location? A crucial importance was the setting up of the make-shift structure that served as the Westville Resiliency Is Us disaster response center. More economically privileged communities enjoy spatial privilege that communities of disadvantage lack. This church had a large multipurpose space that made pitching a large make-shift structure possible. Further investigation, including media reports, revealed that a wealthy Irish contractor who had secured large contracts in the rebuilding of ground zero after 9/11 and would later gain contracts in The Rockaway recovery through the mayor's Rapid Repairs program put up this tent. This remained a mystery to residents and volunteers. All the residents thought was that a wealthy Irish contractor just set up the structure and left. Another source of Westville's local *crisis capital* was Peninsula Circles, which also coordinated fundraising and disaster response that benefited the Westville area.

My interview with Kacie, a volunteer who began working in what would later become the Resiliency Is Us disaster response center only a few days after the storm further explains the cumulative process of how various organizations came to function within this Westville disaster response center:

> KACIE: So, once they set up the tent, then the organizations would just—I was here at the center on the days that it happened. LIPPA [utility company] would come in and say, 'We're here. We can talk to residents. Can we have a table?' And now here I am. Nobody. Just absolutely nobody. Oh, and Monsignor going around, and Monsignor will be like, 'Well, ask her,' and I'm like, 'Ask me?' Is it—'Ask me, why ask me?' And I'm like, 'Okay. LIPPA, you can set up there. FEMA, same thing.' And we—FEMA didn't—FEMA didn't find this center for a month.
>
> SM: How did FEMA finally get to you?
>
> KACIE: That's the point because then they finally started to see the tent, and they started to put their heads in the center and say, 'Can we come here?' 'Of course, you can come here.'

Why would Resiliency Is Us decide to set up in Westville? In deciding where to set up, representatives of this large NGO would drive around to see where there was already activity. They depend on local input as well. My interview with Megan, the Resiliency Is Us's field manager who later took over the response operations in the Westville disaster response center, revealed that there were both push and pull factors. The considerations were beyond assessing which areas were hardest hit.

> SM: Some neighborhoods are visibly destroyed, but some aren't. How do you know where to set up?
>
> MEGAN: Um, you work really closely with community partners like local volunteers and the city. And our volunteers, we won't—I mean, we're human too. If we see that there's not a need for it, we'll report it too.

Here, Megan is saying that the input about where to set up comes from the local volunteers in an impacted area, but that Resiliency Is Us also makes the determination about need.

Once Resiliency Is Us set up operations in Westville, several smaller, less well-known NGOs also came into the area. An important aspect of deciding where to set up the base relates to the presence of a giant in the humanitarian aid industry, such as Resiliency Is Us. For smaller organizations looking to gain legitimacy, and future donor funding, working with a large and reputable NGO is an important endeavor. I spoke with a field manager (who was also a board member) for a smaller NGO working with Resiliency Is Us in Westville. Since this smaller organization was from another state, I asked the field manager how their organization decided where they should set up their operations. She responded, "work orders, regional leaders, mapping and work orders from online applications" from their members, leads from the media as well as where Resiliency Is Us had already established. The field manager of this small NGO stressed the importance of getting to work with Resiliency Is Us. When I asked if her organization had preexisting ties with Resiliency Is Us, she said no, but she conveyed that it was important to work with them. She stated that her organization processed "thousands of volunteers" for Resiliency Is Us and hinted that this would help them with playing similar roles in future disasters.

The kinds of ties formed between organizations differed across neighborhoods. While some large churches and NGOs talked about collaborative ties with Resiliency Is Us, other organization relations were distrustful. Greg, the founder of a local community-based organization in one of the hardest hit areas in Brooklyn, talked about having to conceal from Resiliency Is Us the actual numbers of constituents in his community to gain adequate resources for their area. Similarly, the founder of Always With You in Eastville thought Resiliency Is Us was "rationing" supplies and that it was not interested in collaborating with them. Other interviews with volunteers and residents suggested that the NGO was there to "take over," a claim that Resiliency Is Us volunteers refuted. For local community-based organizations, the ability to create collaborative bonds with "anchor" NGOs such as Resiliency Is Us was an important missing link to resources in the communities. This link of collaborative relations with Resiliency Is Us was crucial to the organizational agglomeration process that occurred in Westville.

Canarsie: A Host Model

The large local church hosted the FEMA disaster response center, where the managers, volunteers, and staff with whom I spoke assisted disaster survivors. This permanent structure differed from the Westville Resiliency Is Us center, a repurposed tent near the large Westville church. At the FEMA disaster response center, there were representatives from the Small Business Administration (SBA), Housing and Urban Development (HUD), and other New York City and State agencies and organizations. In a smaller separate room, there was a distribution center with cleaning supplies and food supplies. In contrast to Westville, where Resiliency Is Us was the dominant out-of-town actor in the response center, in this location FEMA was the dominant out-of-town entity.

In Canarsie, as in Westville, the Catholic church, headed by Father Francis, was the initiator of what would become the Westville Resiliency Is Us disaster response center. Like the Westville church, this center ended up serving thousands of residents. However, the disaster response process was quite different. The organization hosting response building in Canarsie, where the church simply facilitated the state and non-state relief organizations. When I arrived at Father Francis's church, there were several governmental and non-governmental agencies and programs with representatives stationed at various tables throughout a large space. I seized the opportunity to interview Father Francis about the response-building process at this center:

> SM: Can you tell me who you are and how all this came together?
>
> FATHER FRANCIS: Father John Francis . . . and it was gratuitous that we happened to have a building here that could be used for FEMA after the disaster occurred, because our school was not rented. And therefore, or used as a school. So, we were able to invite FEMA here to, uh, our parish, so that the people in the Canarsie area could, uh, have the help that they needed. Because the first few days after the, uh—for a long time really after the hurricane, Canarsie wasn't even recognized as an area that was flooded. They lost so much, they lost everything. And then we had a meeting here in this room.
>
> SM: A meeting with who?
>
> FATHER FRANCIS: A meeting started by, uh, Mr. Perry and Mr. Sampson, the assemblyman and the state senator of the area of New York. And they gathered people here, and the first night we had 1,300 people in this room and in the cafeteria, which is on the other side.
>
> SM: And what date was that?
>
> FATHER FRANCIS: I was here for about ten days. It was before that. A couple days before that.
>
> SM: Before that? Okay.
>
> FATHER FRANCIS: Yeah, just before that, and then FEMA said they would try to

open a place here [in Canarsie]. And then I said, 'Well you've got the place, use this.' The state senator got everybody together, and then they announced there would be a meeting here at this auditorium, because it was the largest place that was free at the moment. And then we got it. Then through the politicians and got FEMA, and we got the people together. We were able to get people to recognize that the need is here, and this would be a great place to have . . . So, FEMA and the church provided it, you know. So, we were there to be able to serve the community here in Canarsie. So that's how things got together.

In this instance, the response building began as a political process where a local politician was able to make all the necessary connections with FEMA as well as with residents who are their constituents. The church was significant because of the spatial resource it was able to provide for the gathering and because it was able to extend the invitation to "host" these governmental entities. Father Francis continued:

FATHER FRANCIS: And then it became not only a place for FEMA, but then became a distribution center also. For the longest time it was a large distribution center.
SM: And how did that start—the distribution?
FATHER FRANCIS: Got started through, through three people from, from the area—from Senator Samson's office, uh, this woman, Valerie, who I don't know if you've met her. She was here with us just now. And Brandon. Brandon works for Senator Sampson. They volunteered to help, and then we were getting all kinds of things from Facebook and all kinds of things people were volunteering. We had one beautiful little thing that happened. We got a big truckload of, uh, nonperishable food in boxes from Porterville, Illinois—Catholic parish there.

Although Father Francis stated at the beginning that the response building was "gratuitous," this process shared some commonalities with the Catholic church in Westville. They both had access to a large "free space" that could be used by FEMA and other organizations, and both resulted in being a "nucleus of relief" in the respective communities. However, there are key differences here also. While elected officials such as the mayor, the state senator, and others participated in getting federal resources to these areas, in Canarsie the elected officials played a significant role in initiating and facilitating the partnership between FEMA, the church, and residents, which led to FEMA setting up a disaster response center. The local government also played a direct role in getting private donations to this location, which other parishes supplemented.

Brooklyn: Organization Coalition

Yet a third model of response building is Organization Coalition. In contrast to the Westville case, the Greater Brooklyn disaster response effort was due to pre-existing organizational and political ties among faith-based and non-faith-based organizations. The presence of hundreds of small and large church partnerships with elected officials prior to the disaster was instrumental in this kind of coalition building. This is due in large part to the high participation of communities in Brooklyn politics. Brooklyn politics, as in many urban areas, is very Caribbeanized due to the high representation of Caribbean American elected officials. Also, many of the church leaders in Brooklyn are from the Caribbean and have been involved in this "social gospel" in responding to the needs of the West Indian community. The fact that the elected representatives are also from small island nations in the Caribbean also helps facilitate what I refer to as a brand of politics that sets expectations for an elevated level of responsiveness from politicians in matters that affect their constituents. These various leaders and their organizations had already been working closely with the local government on issues relevant to the community during routine periods. This means that when the disaster hit, these groups and leaders relied on their existing *social capital infrastructure* to respond to the disaster. I illustrate this process through my interview with Reverend Dennis from the United Methodist Church located in Brooklyn in an area unaffected by the storm.

I attended the second meeting of what transformed into the Brooklyn Long-Term Recovery nonprofit. A large church hosted this coalition. In attendance were about twenty-five organizations and Resiliency Is Us. At that meeting the organizations decided that they needed to form a separate 501(c)(3) organization to petition for federal funds that would become available in the coming months through FEMA grants. All these organizations were already involved in disaster response and recovery efforts but wanted to gain access to federal funds to rebuild Brooklyn. At the close of the meeting, I had a chance to interview Reverend Dennis, the pastor of the church hosting that meeting. I asked him:

>   SM: How did you come together to work on the disaster?
>   REVEREND DENNIS: The meeting today, uh, was formed out of volunteers working in disaster response. So, it's, it's a broad spectrum of interfaith, interreligious, nonreligious community groups. Just people working in

response to the disaster. And trying to mold, uh, an organization, a structure, so that we can respond in a cohesive manner to the, uh, disaster response.

Once more, I was interested in the same basic question of the process of response building. I asked him:

SM: When was the first time that you met as a group?

REVEREND DENNIS: We met about a month ago for the first time as a group in this place—in terms of the groups working in Brooklyn specifically.

SM: Okay. So, who initiated getting the group together?

REVEREND DENNIS: Well, it was through, um, uh, several groups, actually—ourselves, through FEMA.

SM: Did FEMA contact you or you reached out to them?

REVEREND DENNIS: Well, it was through a sort of mutual contact between one of the FEMA VALs—Voluntary Agency Liaison, and myself and also World Circle and several others.

SM: Okay. So you're saying that these were preexisting relationships?

REVEREND DENNIS: Not necessarily so. Although we are a part of NYDIS, which is New York Disaster Interfaith Services. And New York VOAD—Voluntary Organizations Active in Disaster. And, um, so there was some preexisting, uh, relationships, but not in terms of working in Brooklyn—because we haven't had a preexistent group. So, it came out of, uh, just, um, active activity in the Brooklyn area in response to this storm.

SM: Okay, so, um, do you know roughly how many people were at today's meeting?

REVEREND DENNIS: Probably about twenty-five, thereabouts.

SM: And of those, how many of them did you, your organization have a relationship with before the disaster?

REVEREND DENNIS: Well, we've had a relationship with Resiliency Is Us. When you say organization, what are you talking about?

SM: Just organizational relationships. Knowing them or having worked on something with them before.

REVEREND DENNIS: Well, we have, we've had Presbyterian Services and the Lutherans and Presbyterians. We've had Resiliency Is Us. We've had several others. So, they're about maybe a dozen or so we've had relationships in other places with, in terms of the United Methodist Church and UMCOR—United Methodist Committee on Relief—relating to them in, in various places. Like, for example, in response to Katrina, in response to Irene last year, and so on. So, we've had some preexisting relationships on different levels in

Organizational Networks 127

> different places. So, I will say, this office was formed directly in response to the storm.
>
> SM: Right. I hear you. I guess what I'm interested in is, how do organizations come together to respond to a specific need, and whether or not these were preexisting relationships.
>
> REVEREND DENNIS: Okay, okay.
>
> SM: When and how these unfolded.
>
> REVEREND DENNIS: Well, as I said earlier on, we have a preexisting committee structure, based on response to disaster. So that was already in existence.

Having preexisting relationships among organizations meant that meeting organizers announced meetings through preexisting listservs. Many of the leaders of these organizations had worked together on other issues and in some cases had also created interpersonal social capital with others in these organizations on which they were later able to draw during the response to Sandy. The interconnectedness of the politically oriented majority Black and Black immigrant community churches in Brooklyn and other areas also helped facilitate this coalition building. Although brief mentions of elected officials were part of the story, unlike the case of the Canarsie disaster response center, elected officials were not the main actors here. Also, preexisting ties, rather than new organizational ties, were significant in the response building. This coalition among small and large nonprofits and local governments around long-standing issues also proved to be an asset, evident in my conversation with Reverend Dennis. I asked him:

> SM: You're saying that you've not had an issue with resources in terms of transportation, getting generators?
>
> REVEREND DENNIS: To some extent. Well, of course, we always need more equipment.
>
> SM: Where did you get the resources from?
>
> REVEREND DENNIS: We have some equipment and resources of our own. Um, one of the ways that we are able to respond fairly quickly to disasters is through the membership of our church nationally giving. So, we have UMCOR, which we can put in a request for funds for disasters.
>
> SM: And how much have you received just for this local community?
>
> REVEREND DENNIS: For this they haven't specifically said, 'Here is $20,000 for Brooklyn.' What happens is it operates through our conference. So immediately a disaster occurs and, and we have a structure through which it works. The bishop is able to ask UMCOR instantly for $10,000. And there is no red tape to go through. They haven't had to say, 'Well, we need money

for this or that.' And that's our immediate response as a United Methodist Church.

SM: So, you got that? You got $10,000?

REVEREND DENNIS: We got the $10,000 upfront.

Large churches are a significant part of the story in all three cases. Large, long-established, high-capital churches, unlike the low-capital storefront immigrant churches in Eastville, have a built-in advantage in responding to disasters. First, they tend to have several branches dispersed throughout the country. This means that at one point or another, they will have gained experience responding to emergency events and can quickly mobilize their resources. These churches can quickly access funding through their headquarters or conferences. Another advantage of these multibranch large churches is that if one location experiences a disaster, unaffected locations in other cities and states can help both in terms of labor and finances. They are also able to attract and deploy volunteer members who do not live in disaster-impacted areas. Even for their affected members, this serves as an "entry point" into the communities to know exactly where the need is.

All three of these models of *response building* among disaster response organizations seen in Westville, Canarsie, Brooklyn, and Eastville bore out characteristics of the urban areas institutional and organizational environment, the spatial resources of large organizations, and the community social capital potential in the form of financial and political capital of its current and former residents. In Westville and Canarsie, having available space through the presence of high-capital branch churches such as Catholic and Methodist churches was instrumental in attracting attention and pertinent actors. Westville's Irish ethnic enclave and connections to affluence led to instituting the tent. Similarly, local organizations already having considerable command of pre-disaster resources helped them mobilize instantly.

The coalition of small and large local organizations in Brooklyn positioned themselves to become an instant source of social capital. The political and organizational interconnectedness with large NGOs and federal funding gave them a steady stream of access to pertinent information and resources. This means that they could quickly repurpose their ties toward securing disaster response and recovery funding not just for Superstorm Sandy, but for future disasters. This coalition model gives small Black and immigrant churches access to organizational social capital they would not have otherwise had. Each model presented quite different opportunities and constraints on organization-mediated disaster response, the degree of accessibility of resources, the timing of information spread, and the kinds of disaster experiences of residents across these urban disaster areas discussed in this book.

Eastville: Organizational Isolation

This most vulnerable residential area in Eastville lacks the social, spatial, and institutional infrastructure that would enable Always With You to serve as a host for local and national organizations or to build coalitions and attract lead NGOs to set up their bases there. In the disaster response area near the Eastville Always With You, there is not this interconnectedness of high-capital large churches and low-capital small churches with high-capital governmental organizations. Eastville also does not have the political mobilization we see in majority Black areas in Brooklyn. Eastvillers live in an urban area that has suffered decades of institutional disinvestment, high political disenfranchisement of its residents, and very few organizations outside social services serving the needs of economically deprived urban residents. The low-capital community-based organization Always With You, like the immigrant-serving church in this area, had lacked spatial resources since it operated out of a small storefront space.

In the most vulnerable part of Eastville, many of the residents are the most marginalized economically deprived, which includes formerly incarcerated and substance-dependent persons. Not only are they geographically and politically disconnected from the inland areas of the city, but they are also hyper-segregated from economically privileged areas such as Westville. Therefore, their *crisis capital* does not become augmented by collaborative relations with governmental and nongovernmental organizations in the way that the three models described above do, albeit they did have engagement with some of these organizations. For example, in Eastville, FEMA applications at one point were located in a grocery store, rather than the mutually reinforcing space of the Canarsie Catholic Church.

Similarly, urban areas such as Eastville are not able to benefit from *organizational agglomeration* that would in turn attract a *nucleus of relief* around a large high-capital NGO as are urban areas such as Westville. In Eastville, the transactional relations between Resiliency Is Us and Always With You are quite limited and sporadic. For example, Always With You expresses that Resiliency Is Us delivered bottles of water to them but left shortly thereafter. Another time they came with a mobile clinic but did not hand out any medical supplies in Eastville. Also, when Always With You did receive a donation of their own mobile clinic, Resiliency Is Us did not assist with sharing medical supplies needed to keep the clinic running.

Unlike the participation of Resiliency Is Us in the Brooklyn long-term recovery coalition meetings, Always With You volunteers recall attending community meetings in Eastville with other local community-based organizations and Resiliency Is Us being "notoriously absent." In short, the most vulnerable part of Eastville's *response building* was stymied due to what I am calling an *organization isolation model* of response. This type of response characterized an absence of a spatialized, collaborative relational field with high-capital governmental and nongovernmental organizations organized around disaster response and recovery.

CONCLUSION

Ecologies of Inequity in Disaster Response

> There is a chasm between human suffering and institutional support during disaster response. We need to make visible how higher-order racializing and classing logics inscribed in segregated urban spaces, organizational practices, and interpersonal relations unleash exclusionary biases, blind spots, and behaviors to deny equitable access to the poor and racially minoritized disaster survivors. We need to learn about the making of inequality, so that we can figure out a way to unmake it.
>
> —Sancha Doxilly Medwinter, the author

Post-Sandy Rockaways and Canarsie present many lessons learned about race- and class-differentiated communities, their relationships to governmental and nongovernmental organizations, and their access to timely information and resources pertinent to recovery outcomes. Superstorm Sandy has taught us that the configurations of FEMA, NGOs, and NYS disaster response centers help create an *ecology of inequity* through racializing/racialized and classing/classed urban spaces, organizations, and urban residents. An ecology of inequity bestows ecological privilege to one community while relegating ecological deprivation to another. In the former scenario, a robust configuration of networking organizations and responders produces social capital relations that transmit unique access to resources and information. This concentration of organizational networks and interpersonal relations benefited disaster survivors in Westville, the White economically privileged urban area. On the other hand, the absence of this concentration in economically deprived urban areas relegates an ecology of deprivation in places such as Eastville, excluding disaster survivors in this area.

NGOs Home-Field Advantage Creation Excludes the Vulnerable

The locational decision of Resiliency Is Us, the large NGO, helped Westville become better bridged to high-capital NGOs. High-capital NGOs are, in turn, better linked to governmental resources (Woolcock 1998; Woolcock and Narrayan 2000). The convenient location of Resiliency Is Us also incubated organizational, community, and interpersonal social capital. Westvillers reaped the benefits of hypervisibility of their suffering and their disaster work, the opportunity to become "regulars" of the NGO center, and, consequently, the relative advantage of frequent interactions with NGO responders.

Westville's interactional context allowed Westvillers to then build affective, emotional bonds with NGO responders, the brokers of the high-capital NGO. As a result, Westvillers were able to influence the on-the-ground decisions of Resiliency Is Us and reciprocally receive and share continual, informally transferred, time-sensitive disaster information and resources. On the other hand, Eastville, the economically deprived community adjacent to Westville, remained starved of such a lucrative ecology of structures and relations and the informational and resource exchanges they transmit.

The "home field advantage" that the Westville center provided to Westvillers enabled them to engage in mechanisms of social closure, quasi-privatization of public space, and hoarding of public goods. Westvillers also engaged in exclusionary narratives to substantiate notions of who belonged to Westville and therefore had a right to access disaster resources. One example is determining that people who looked clean, who were coincidentally non-White, could not be Sandy survivors and therefore were not entitled to the disaster supplies.

Always With You, the Eastville community-based organization, due to its limited resources primarily provided Eastvillers with *bonding* social capital, the expressive type, punctuated with short stints of fleeting crisis capital. Always With You was spatially and organizationally isolated from Westville and Resiliency Is Us, and it severely lacked its own resources. Even the location of Resiliency Is Us on the periphery of Eastville was dormant primarily due to the NGO's strained relations with grassroots community volunteers, who were already serving the community before the NGO's arrival. Resiliency Is Us also interrupted the early mobilization of bonding social capital around what would become the NGO's Eastville

Conclusion 133

location. Yet, the NGO was ill-equipped to replicate the "warmth" that the impoverished, racially minoritized, White, and new immigrant disaster survivors had received from members of their community before the NGO's arrival.

Eastvillers had long-standing bonding social capital with the founder and volunteers of Always With You. However, these community-based organization responders did not have the needed links to institutionalized resources and information that was available to NGO responders through the NGO's unique access to federal resources. There was also no opportunity for the Eastvillers who trekked to the Westville NGO center to forge bonding social capital with the NGO responders, the custodians of institutionalized resources, due to their embeddedness within the high-capital NGO. The founder and volunteers of Always With You repeatedly expressed a desire for a sustained, collaborative relationship with Resiliency Is Us, but the most they received was the occasional drop-off of an inadequate number of basic supplies such as cases of bottled water. This situation of misalignment of resources away from the most vulnerable communities they served was incredibly vexing among the Always With You responders, as well as other community-based organizations and small churches in Brooklyn.

NGO Organization Agglomeration Exponentiates Inequity

When an NGO comes into an urban disaster area from outside and selects a location that is most convenient to White economically privileged survivors, this is a loss of potential material and informational resources for adjacent economically deprived, marginalized communities. This loss is in the form of lost opportunity for the latter to also benefit from "home field advantage," have influence on the on-the-ground implementation of disaster response, and receive timely, pertinent information to secure institutional resources.

However, the magnitude of loss to the urban economically deprived community is not merely the loss of what one organization provides. When an NGO that commands influence, resources, and connections with governmental collaboration locates near a White economically privileged urban area, the distributional inequities are exponential. This locational decision amplifies inequity because large and small NGOs in the disaster response and recovery organizational field may view close proximity to this NGO as allowing them to gain reputational status and organizational capital. This dynamic was evident in my conversation with the field manager of a small NGO who revealed her hope that providing support services for Resiliency Is Us would position her organization for future opportunities to collaborate with the large NGO in its international disaster response missions.

FEMA's Bureaucratic Logics Displace the Vulnerable

Governmental organizations also play an instrumental role in the creation of *ecologies of inequity*. They do this by transferring to NGOs the huge responsibility of brokering State resources. Their placement in buildings with local governmental and nongovernmental organizations, and high-resource churches also contributes to organizational agglomeration. A third way that they contribute to an ecology of inequity is through the *logic of response* carried out by disaster response organizations and their staff and volunteers. This logic of response comprises the practices, assumptions, and expectations of governmental organizations and responders. This disaster response logic reflects and caters to White and middle-class privileged statuses, while it disadvantages and makes invisible the hardships that are unique to those who are non-White, lack legal status, and are economically deprived.

The logic of response displaces the centrality of the *logic of services* in economically deprived communities. The logic of services is what typically sustains, albeit ineffectively and inadequately, vulnerable populations in economically deprived urban areas. The prioritization of disaster response displaces the prioritization of the routine needs of those who have always experienced institutional neglect and economic deprivation, prioritizing the recent and immediate needs of "disaster victims." However, this category does not distinguish those who are only temporarily experiencing acute deprivation from those whose acute deprivation compounds with their decades-long, in some cases, chronic deprivation.

The logic of response is race and class blind. As this study illustrates, FEMA-run centers and responders erased significant distinctions of deprivation that causes vulnerability among survivors. The logic of response of the ecology of inequity is also not as responsive to the disparities among homeowners and renters. This logic is especially blind to the unique plight of basement renters, who had to watch their entire life's work be submerged under flooding that reached their ceilings, as they escaped with nothing. This logic renders basement renters invisible because this class of renters does not "legally" exist, which means that many undocumented experience yet a second type of legally defined invisibility and exclusion.

This logic of response promotes a *middle-class bootstrap bias* that more easily recognizes and recovers losses to the formally employed than to the self-employed. This logic of response must not fully comprehend the value of the tools, nor the value of the DJs, the peddlers, and the backyard mechanics, nor the value of these urban economies of survival.

This logic of response also does not sufficiently consider how the placements of disaster response organizations combine with the fragmentation, depletion, and deflation of social ties of disaster survivors. Furthermore, it does not see how the cultural rules around asking among racially minoritized and impoverished populations further compromises the utility of traditional networks. This logic is also blind to the fact that the networks of the racially minoritized and economically deprived urban residents suffer the most during disasters. The fact that the networks of the racially minoritized tend to be concentrated in disaster areas is not insignificant, because this means that social ties that are the key to navigating disaster assistance are themselves multiply displaced and dispossessed.

Another tendency in this logic of response is to reward "early birds" and penalize latecomers who are seeking disaster assistance. This logic of response does not fully comprehend the confusion around how legal status may impact FEMA grant eligibility or how FEMA's relationship to the Department of Homeland Security might discourage some disaster survivors who are part of mixed-status families from seeking disaster assistance at FEMA-run disaster response centers.

This logic of response has not yet fully grasped the urgency to assuage the challenges governmental organizational support pose to the chronically economically deprived and substance-dependent program recipients. The logic of response does not see that they have been cycling through transitional housing and homelessness for years under the logic of services to then again be offered a voucher for yet more transitional housing, again with no clear path to permanent housing. The logic of response through these various pathways transmits inequity like race, class, immigrant status, and poverty status.

Centering Race, Class, and Social Capital

In discussing the specific policy implications of what Sandy has taught us, I use as a baseline the policy recommendations that FEMA already has at its disposal. On November 20, 2020, the National Advisory Council (NAC) presented Administrator Peter Gaynor a report featuring the equity issues, among other things, with FEMA's execution of disaster response. The NAC 2020 report cites FEMA as not meeting the civil rights requirement of the Stafford Act, which explicitly states the requirement of racial equity. The executive summary of the 2020 NAC report confirms what I had already learned while in the field: that FEMA's disaster response "provides an additional boost to wealthy homeowners and others with less need, while lower-income individuals and others sink further into poverty after disasters" (NAC 2020, Executive Summary, 6).

The NAC report also identified issues of equity at the heart of the overlapping, disproportionate, negative impacts of Covid-19 on communities that are already "socioeconomically marginalized" (NAC 2020, 7). The NAC report's glossary defines its equity goal as "everyone meeting a minimum outcome." This would mean that Ricky, who had slipped into a deeper stage of homelessness after Sandy, "would not be homeless" if FEMA implements this needed corrective (50). These recommendations are good first steps, although the benchmarks leading to full implementation by 2045 remain unclear at the moment.

It is important to note, however, that neither FEMA nor the 2020 NAC recommendations explicitly deal with the question of how racializing and classing structural and relational processes connect to produce inequity in disaster response. For example, where the report raises the subject of equity, we do not find any mention of race or racism (NAC 2020, 7–8). In the whole report there are two mentions of race that are part of a list, which appears in a quote from disaster scholarship and another in quoted language from the Stafford Act. The word "racial" appears twice, both in reference to racial diversity. There is zero mention of racism, neither institutional, environmental, nor interpersonal. Furthermore, the report connects the need for equity to a vague statement on "nationwide protests and unrest" and "social disruption" (7–8). This stops short of a clear articulation of the Black Lives Matter movement as an explicit, collective cry for racial equity and racial justice, of which disaster justice, with race and class at its center, needs to be a part.

The NAC 2020 report situates social capital within a framework for achieving equity, citing social capital as the "main contributor to the effective recovery of a community post-disaster" (45). The report relies on a conceptualization of social capital that emphasizes "norms," "trust," and "networks" (15) and is operationalized at the community level. The report also states that communities with low social capital experience low recovery, while those with high social capital experience high recovery. This social capital framework leads to a focus on building trust between communities and emergency management systems.

While the above definition of social capital has its utility, this Sandy study's conceptualization of social capital emphasizes the information and resource conferred from high- to low-capital social ties at the organization-to-organization and person-to-person (dyadic) levels. More importantly, this study finds that race and class structures and processes impede the access and mobilization of pre-disaster social capital and therefore their utility during disaster response. Race and class structures and processes simultaneously interrupt the social capital creation for economically deprived areas and their residents, while facilitating this process for economically privileged areas and their residents. These findings point to a need to systematically investigate and address the race and class dynamics that hinder equitable networking opportunities of disaster survivors and community-based organizations to produce social capital during the disaster response period.

Conclusion

This approach conceptualizes and operationalizes social capital at levels that make interventions immediate and feasible. Adopting this approach also avoids the slippery slope of resigning whole communities to an uncritical fait accompli explanation that says, poor investments in pre-disaster social capital, then low recovery outcomes. From an implementation standpoint, there is also a danger of resorting to simply summoning social capital explanations as a proxy for serious interrogations of how and why ecological racism and classism leads to inequitable social capital creation opportunities during official, organization-mediated disaster response. As this study has shown, the relational processes of race and class interrupt the relational processes of social capital at multiple stages and levels. Therefore, to the extent that we care about social capital as it relates to equity, we need to keep in focus race and class structures and processes.

The NAC report also stresses the need for FEMA to place an "emphasis on local capacity" because the multistate approach has become financially prohibitive, given the frequency and intensity of disasters as well as myriad supply chain issues (NAC 2020, 45). Beyond the reasoning provided, this Sandy research identifies additional drawbacks of a heavy reliance on nonlocal response. One of these drawbacks is the lack of urban cultural understanding of nonlocal responders, which inhibits FEMA responders' understanding of the way that urban residents structure their lives around local city living. For example, Beverly, the FEMA staff person from a midwestern state, did not understand that living in a basement apartment in New York City is not as peculiar as she perceived.

Similarly, people who are racially, ethnically, and culturally different from disaster survivors and who have little experience assisting racially and socioeconomically divergent populations do not understand the particular needs of these urban disaster survivors. They do not possess the racial, ethnic, and cultural understanding and the keen awareness of how race and class structures and processes create and maintain urban inequality. Their implicit assumptions and biases reduce the capacity of these responders to deliver assistance equitably. This problem became evident in the unfortunate slogan of a FEMA site manager who confidently stated "First you get the needy, then you get the greedy" in response to my question about a noticeable lag in disaster survivors coming in to seek disaster assistance.

Both the NAC report and FEMA seem to converge on culturally sensitive training of FEMA employees to address biases and assumptions. I agree, and I return to my recommendation of hiring local residents, who are already organically endowed with a thorough understanding of the challenges of navigating the unique terrain of the city. Disaster responders need to reflect racial, ethnic, and local community membership of the disaster survivors seeking assistance.

Implications for FEMA Response

Specifically, FEMA will need to significantly reduce its bureaucratic burdens. One primary avenue to achieve this is to streamline the labyrinth process of applying for grants. The process needs to be more linear with less emphasis on obtaining denials and submitting appeals. Many Superstorm Sandy disaster survivors with whom I spoke both in Canarsie and The Rockaways incurred what Pamela Herd and Donald P. Moynihan (2019) call "administrative burdens." Administrative burdens are the unnecessary costs (i.e., learning, psychological, and compliance) that citizens who need to access a public benefit from the government have to incur in order to receive such benefit. FEMA processes unequally distribute administrative burdens, as not everyone necessarily has the material, human, and social capital to gather information to apply for benefits.

In particular, there is a huge tax on the emotional and psychological bandwidth of the racially minoritized and economically deprived disaster survivors when they navigate FEMA's labyrinth appeals process. Finally, there is a prohibitive cost of compliance to administrative burdens (Herd and Moynihan 2019) such as the financial cost of applying for a SBA loan or the psychological cost of making multiple trips to centers, finding paperwork, and contacting third parties. Some disaster survivors found these burdens insurmountable and gave up before receiving any benefit. This cost is high even for those who dare to stick with the process to the end. The psychological cost of administrative burdens was particularly visceral among the racially minoritized, noncitizen, elderly, and economically deprived mothers and fathers, basement renters, and the multiply displaced and homeless in Canarsie and Eastville. Contrastingly, there was a marked absence of this anguish and fatigue among Westvillers, who often described their FEMA application process as smooth and uneventful.

FEMA needs to reallocate significant aspects of disaster response as the responsibility of governmental organizations. The staggering human suffering, deprivation, and psychological burden falls to private organizations such as churches, community-based organizations, and, to a lesser extent, NGOs. Where the State has failed, these organizations have tried to fill in. However, while nonlocal NGOs have unique access to institutionalized resources, they are not best suited to provide disaster response in impoverished and marginalized communities. We see this when we compare and contrast the disaster response pitfalls of Resiliency Is Us with that of Always With You. While local churches and community-based organizations take on this huge responsibility of meeting the needs of the economically deprived, they provide services with little to no funding. Therefore, I suggest a tighter partnership with governmental and local community-based organizations, where the former provides a steady stream of information and resources as well as organizational capacities to the latter, while the latter continues to provide the person-to-person, block-by-block assistance to their communities that they are the most equipped to provide.

FEMA needs to place less emphasis on a compliance-based, fraud-detecting model of disaster response and place greater emphasis on a deprivation-cognizant model. FEMA uses subcontractors to deliver services, such as nonlocal adjusters. These nonlocal adjusters focus on detecting disaster fraud (GAO 2015). Unfortunately, unsuspecting disaster survivors are confused when they receive a denial letter. The Sandy survivors did not understand why adjusters had taken numerous photos or viewed the photos that disaster survivors had taken themselves yet did not include those in the claim. Some Caribbean disaster survivors followed West Indian cultural norms that stress keeping a clean home. Therefore they had quickly cleaned up the filth from their basements in time for the early pickup by the sanitation trucks. Unfortunately, this meant that by the time the adjuster arrived, much of their evidence was gone. Since then, a fact sheet updated by FEMA in March 2021 instructs that disaster survivors should not wait on the adjuster before cleaning. However, it is unclear whether adjuster investigative methods have significantly changed to align with these guidelines.

Some Canarsie disaster survivors described their impressions of adjusters' assessments as seemingly arbitrary, lacking empathy, and conveying condescending remarks, as if their personal belongings lacked value and did not need replacing. Some also thought that the process was inequitable, especially those who had compared grant amounts with their neighbors who had incurred similar damage. Furthermore, the feeling that these unfavorable assessments were related to racial stereotyping led to even more frustration and despondence among these disaster survivors.

FEMA needs to examine and address how its on-the-ground operations alongside other organizations may usher in a logic of response, which displaces the logic of services. Addressing its contribution to the Logic of Response will be indispensable for future response to communities marked by heavy reliance on social services. It is paramount that FEMA works seamlessly to integrate social services and disaster response. This would reduce the possibility of exacerbating the suffering of the chronically economically deprived. This population depends on social services in routine periods, which only increases during and after disasters.

Related to the earlier points about the need to transition away from nonlocal responders, local social workers with extensive social services experience need to replace nonlocal FEMA field and site managers. These local professionals are best suited to manage disaster response centers, precisely because they customarily work within a framework that sees and understands long-standing trauma and deprivation. This recommendation comes from my extensive interview and observation of how site manager Caroline runs her disaster response center.

When compared with how nonlocal FEMA site managers run their sites, having a local site manager with a social services background brings indispensable value to disaster response. Caroline, who had twenty years in social services, explained that providing social services cannot be a mere afterthought, secondary to a preoccupation with disaster response. Drawing on her model of enlisting the services of a local mental health clinic, I recommend the presence of on-the-spot mental health counseling in all disaster response centers. The current practice of referrals to medical professionals only further strains the thin psychological bandwidth of the most marginalized. Furthermore, disaster survivors may have already been struggling with undiagnosed and untreated mental health challenges before the disaster event. This means that these disaster survivors are experiencing a further compounding effect from the trauma of experiencing the disaster event and having to navigate FEMA's labyrinthine grant appeals process, among other disruptions in their lives.

Implications for NGO Response

NGOs need to broaden their on-the-ground implementation goals beyond merely trying to bring disaster supplies to the general vicinity of economically deprived impacted urban areas. Large, reputable, nonlocal NGOs need to set and follow through with explicit, specific objectives of establishing their base of operations in the most economically deprived and dense residential urban areas. Their consistent, visible presence in these economically deprived disaster areas would reassure these habitually neglected urban residents that their needs are central to disaster response efforts. By taking this step, they will provide the possibility for greater and more timely access to informally transmitted, up-to-date information and resources.

Pursuing and fulfilling this goal will incubate social capital in economically deprived areas through NGO responder relations with community-based organizations, as well as among disaster survivors who can comfortably frequent the NGO disaster response center. Additionally, as this Sandy study finds, the presence of a large reputable NGO serves as a magnet for other smaller organizations, leading to the pooling of responders and services to the area. Up to this point, this is the process already uniquely enjoyed in economically privileged, White areas, such as Westville, to the exclusion of racially and ethnically diverse and economically deprived areas, such as Eastville.

NGO responders deploying to an urban disaster area need to have a thorough understanding that urban space is segregated; racial and ethnic groups are hierarchically ranked in terms of wealth, status, power, and influence; race and ethnic relations become animated over scarce resources, to include disaster response resources; and urban space is contested as boundaries and networks become reified to exclude those presumed to be undeserving or not belonging. More importantly, NGO responders need to be cognizant of how these dynamics individually and collectively are set in motion by, and infiltrate, the decisions of NGOs, beginning from where they decide to establish their presence.

This consciousness of race and class dynamics equips NGO responders to avoid the blind acceptance of race- and class-neutral assessments of "safe" versus "unsafe" urban areas. NGO site managers need to become more aware that the decision and action of placing an NGO site in a White middle-class urban area ensures inequity for its adjacent economically deprived communities. Even when they place a satellite location in the adjacent community, they need to ensure that this location is nearest to the densest residential clusters where there is the highest concentration of vulnerable, impacted disaster survivors.

Furthermore, it is insufficient to simply outfit NGO disaster response centers located in economically deprived urban disaster areas like those in economically privileged areas. NGO site managers need to be equally present at both sites. NGO managers and other decision makers need to also consider whether a proposed site placement will be in an area that experiences overpolicing. They need to be aware that establishing their base near a police precinct office may deter disaster survivors seeking resources, especially if the carceral state has, at one point or another, swept them into its dragnet.

In talking with disaster survivors in areas where Resiliency Is Us, the large NGO, had a mobile operation, disaster survivors had often missed or were not even aware that the NGO trucks were dropping off supplies. There were also differences in the resources across communities, such as hot plates of food versus pastries and packaged dry food. Disaster survivors and local responders of small churches and community-based organizations both in Canarsie and Eastville attributed these inequities to an assignment of racial and class inferiority to their communities. In both cases they were aware of the better resources that Westville received. My observations also confirmed this inequity. In both Canarsie and Eastville, the local ministers and community-based organization volunteers understood this disparate receipt of disaster assistance by Resiliency Is Us as racism and classism.

Nonlocal NGOs that are going in to provide disaster assistance to economically deprived urban areas also need to prioritize achieving amicable relations with these communities. Local volunteers and community-based organizations are essential to the survival of these communities even in the absence of disasters. This is why it is crucial that NGOs not disrupt the organic, altruistic relations of communities. NGOs need to adopt a disposition of simply aiding existing, local response with their resources and their logistical capabilities.

In the case of Westville, a community volunteer described Resiliency Is Us as maintaining amicable relations that allowed her to continue to play a vital role in the center. Contrastingly, in Eastville, another community volunteer experienced the same NGO as taking over, disregarding the decisions of the community, and significantly altering how they organized aid. The local Eastville volunteer also attributed the lack of engagement by disaster survivors in this location to these less than amicable relations with Resiliency Is Us.

Nongovernmental organizations and their responders need to be cognizant of how the notion of acute collective crisis, coupled with the visibility of widespread destruction of neighborhoods, masks preexisting spatial, symbolic, and material distinctions of urban spaces and disaster survivors. This vigilance leads to greater awareness of how these racial structures and meanings deploy in the interactional environment of disaster response areas.

NGOs need to be aware that their stable presence in residential areas of White, presumably "safer" urban residential areas gives these disaster survivors an unfair "home field advantage" over their non-White and economically deprived neighbors in adjacent areas. The resulting interactional space encourages territorial language and hoarding behaviors that exclude disaster survivors from adjacent urban areas who are coming in to seek disaster assistance.

In conclusion, those who administer and manage governmental and nongovernmental disaster response organizations need to be cognizant of how *already* racialized and classed urban spaces combine with institutional, organizational, and interpersonal biases and blind spots that burden and render invisible those disaster survivors already marginalized by race, ethnicity, class, and legal status. Furthermore, there needs to be a reorganization of the goals of disaster response toward reducing the interruption of social capital creation, access, and mobilization during disaster response. As part of this initiative, decision makers need to recognize and immediately address race and class exclusionary relational dynamics that exclude the most marginalized and economically deprived urban disaster survivors.

If anything, Superstorm Sandy has shown us why we cannot sidestep the centrality of race and class in our attempt to address the inequities of disaster response. Furthermore, the impetus for addressing inequity in disaster response needs to focus on understanding and addressing how race and class structures and processes create and maintain disproportionate suffering of racially, socioeconomically, and legally marginalized disaster survivors before, during, and after disasters. In the absence of addressing racialized and classed inequities, disasters will continue to reproduce race and class inequality and urban poverty.

EPILOGUE

Nine Years Later

I am thinking it's almost nine years since we first stood shoulder to shoulder handing out disaster supplies on a street corner in Canarsie after Superstorm Sandy, when I make the call to Bishop Fabian. I am finally ready to publish my book and wanted to see how the themes and analyses in my chapters held up after a decade. He tells me that long-term recovery is still ongoing in Canarsie and other coastal areas for homeowners who have not received financial support for the damages incurred since Sandy. He explained the uneven pattern of redevelopment that occurs after a disaster:

> FEMA provided resources and money to rebuild New York City and State, strengthen places to the East River, but homeowners here have not been disbursed. The technicality was that they did not have "flood insurance."

The fact that most homes in Canarsie did not have insurance because they weren't considered to be in a flood-prone zone was something I had encountered during my fieldwork.

I inquire about what happened with the long-term recovery group to which he and other pastors belonged. I was curious to find out whether the small churches that were part of this large heterogeneous organizational network of the long-term recovery group would actually benefit from social capital in the form of funding. Bishop Fabian explained that the group accessed FEMA funds, but that "big business" and larger, higher-capital religious organizations, with established community programs such as daycare centers, were the real beneficiaries. Small, low-capital churches had some influence in how money would be disbursed but could not directly benefit. He also explained the instability of this network. Small churches slowly disengaged because the process of getting funding even for other organizations in the group was also a long one. He reflected:

> FEMA did a good thing. FEMA wanted the community's input. The leaders of small churches would have a say in how money was disbursed. Small churches did not get resources. Except those with community programs like a daycare, but conducting disaster relief did not qualify you. What happened is unfortunately what happens all the time. The process is so drawn out that people become frustrated and move on to something else. Big businesses ended up getting something out of it, and small community organizations did not get anything.

Bishop Fabian further explained that FEMA handed off the responsibility of directly liaising with the group to New York State. This move introduced a bureaucratic burden that these leaders of small organizations could not endure, particularly since there was no economic incentive.

> FEMA thought it was difficult to work with the long-term recovery group since there were several organizations, so it went to the state. The state process became very tedious for members, and they dropped off one by one. So, 90 percent of big businesses recovered. But only businesses stayed on.

In this first scenario Bishop Fabian described, the small churches don't benefit. However, then he described a turn of events, where the leaders of the small churches that met through the long-term recovery group established a separative coalition network in order to position themselves to capture funding first from the mayor and then eventually from the federal government:

> The outgrowth of Sandy and meeting at the long-term recovery group is that small churches have come together to organize. Seven, eight, or nine churches have come together to form a coalition. This was helpful because it allowed us to qualify for the federal stimulus during Covid. It's a gun violence coalition of faith leaders. We work with the police precinct on anti-violence initiatives, but we do more than that. During Covid we gave out food and PPE [personal protective equipment]. Six churches were vaccination sites. We recently got a commitment for funding to work with small churches who are working on this issue in the community. We got to know each other at the long-term recovery group after Sandy. That was when we recognized that if we don't work as a unit, we will not get funding.

Organizational coalition networks among Black and West Indian churches in Brooklyn is not entirely surprising. The prominent level of political enfranchisement among Black and West Indian churches makes Brooklyn a fruitful ground for successful coalition building as described in earlier chapters in the context of Sandy. The challenge for this nonprofit is that the fact the governmental grants establish a dependent relationship with the state can undermine collective efficacy (Vargas 2019). Receiving state funding also risks alienating certain parts of the community with a different vision of advocacy on the issue of violence (Vargas 2019). For example, the way that NYPD defines violence would differ from how antiracist, urban movements define violence. The latter's definition encompasses police brutality against Black, Brown, and Indigenous bodies, which the former's narrowly focused campaigns against gun violence decenter.

Immediately after my call with Bishop Fabian, I called Freddie, the founder of Always With You. I let him know that I was working on publishing the book since we last spoke, and that I was calling to see if my analysis of Always With You still captured their current reality, nine years post Sandy. I learned from Freddie that he had to dissolve Always With You due to insufficient funding to operate. He told me he turned over the small storefront space out of which he operated to a small church. When I asked about the storefront church, he said they also had to give up the space.

However, Freddie mentioned that he is only a few blocks from his old location and that he had founded another organization for "serving marginalized subpopulations in this area," adding, "because they have always been near and dear to my heart." I asked him if this organization did the job training that Always With You did, and he said, "I'm here doing the same thing." He was a bit frustrated during our call, although he was happy to hear from me after several years. He said he was trying to secure a Covid-related loan, but he kept "getting the runaround from SBA."

He mentioned that he is affiliated with Greenpeace International and was assisted by Occupy Sandy, but he did not have any local connections to local organizations. He mentioned wanting microgrids installed on The Rockaways. This would reduce the peninsula's dependence on, and vulnerability to, the utility company's poor management of the electric grid system, which was a major issue during Sandy. I also wanted to confirm that I had not missed that he was ever part of a long-term recovery group like the one in Brooklyn. He confirmed that he has never belonged to such a group.

So now we're left with a question: How do we get large NGOs and government organizations to recognize the ways their disaster response practices contribute to increased inequality between urban economically deprived communities and more affluent, often adjacent, communities, and to change those practices so that they identify and support local community-based organizations operating in urban economically deprived areas to survive and thrive before and after disasters?

It is infuriating that community-based organizations, like Always With You, are so committed to their communities and are best suited to provide the "warmth" that so many Sandy disaster survivors articulated they needed but weren't able to receive from governmental and nongovernmental organizations, yet the daily survival of these organizations has to be negotiated each day because they are starved for capital.

APPENDIX A

Interview Guide
Superstorm Sandy Disaster Survivors

Disaster and Disaster Assistance Experiences
- How were you impacted by the storm?
- How did you know about Resiliency Is Us/FEMA organizations or any other group?
- How far do you live from here? How did you get here? How often do you come?
- Did you have any problems with getting to the location?
- Is there anyone you know or know by name in the disaster response center?
- Do you talk with the volunteers? What do you talk about? Do the volunteers talk with others?
- Have you or anyone you know gotten help, advice, or emotional support from the volunteers?
- Is this a comfortable space for you?
- In situations of natural disasters, we want to make sure everyone regardless of race, immigrant status, etc., gets the same help. Did you think any of these things affected you getting help?
- Do you think people from other neighborhoods had a better or worse experience getting help?
- Are you worried about long-term help?
- Did you see/feel that people were more helpful to each other than normal?
- Do you think race or immigrant status affected who helped who?
- Have you noticed groups that do not usually get along, cooperate, or share information or supplies?
- Do you think in society some people get more attention/help than others

in general? If so, who is at the top/middle/bottom in terms of attention in general? Are things different or similar in this neighborhood?
- Now after the storm, who is at the top/middle/bottom? Why do you think so?
- Have you seen or heard anything that made you think so or do you just have a hunch about that?
- As I try to understand people's experience with finding help, is there anything I have not asked?

Demographic, Economic, Spatial and Housing Characteristics
- Are you currently unemployed?
- What job do you do?
- What is your highest level of education? High school, some college, college degree?
- How many children usually live in your home?
- What are their ages?
- To help me know what neighborhood you live in, what are the cross streets? Zip codes?
- What is your housing type? Apartment, single-room occupancy, private house, public housing?
- What floor do you occupy? Basement, first, second? Do you rent or own?

Connections to U.S. and Metropolitan Area
- Migration is a big part of New York's history. In your family who migrated? Me, parent, grandparent?
- What country? Year?
- Under what category did you/they migrate? Farm worker, nurse, teacher, student, other?
- Are you a spouse or child of a U.S. citizen?
- What is your closest family link to a U.S. citizen?

Connections and Attachment to Devastated Community
- How many years have you lived in this community? At this address? In NYC?
- Will you stay/leave the area? Current dwelling? Why/why not?
- In the community, how many close friends do you have? Close relatives? Neighbors you talk to?

Subjective Valuation of Losses (Sentimental, Economic)
- What hurt you the most when you realized it was gone? Tell me about the items you lost.

Appendix A

- How was the basement used? Residence, storage, recreation?
- If you could place a dollar amount, about how much would you say?

Displacement Experience and Support

- Did anyone in the building have to leave? Yes/no, who?
- How many children/adults?
- Where did you/they go? Neighbor, relative, close friend, hotel, shelter, other?
- Did you/they relocate? Another town, same street, other? Why?
- Did you/they return? Did you (also) evacuate before the storm? Why?

Support Received and Race and Class of Activated Social Ties

- Has anyone reached out to offer financial, emotional support, or give information about disaster relief or recovery?
- Specifically what help did they provide and how helpful was it? Did you first ask for this help?
- Was it someone you have known for a while or someone you met through the storm?
- Do you find the volunteers, workers, managers helpful? In what way?
- How do you know this person/s? What are the personal characteristics of the person/persons? Race, gender, occupation, lives in or out of neighborhood?

Racial and Class Heterogeneity of Social Network

- Do you know anyone who is a_____ (listed profession)?
- Is this person a relative, close friend, church member, neighbor,
- NYC resident, gender, new immigrants, Black/Hispanic/Latinx?
- **Professions**: Nurse, writer, farmer, lawyer, middle school teacher, full-time babysitter, janitor, company personnel, CEO of a big company, policeman, hairdresser, bookkeeper, security, production manager, operator, congressman, taxi driver, hotel bell boy, admin assistant in a large company, receptionist, computer programmer?

Disaster Responders: Volunteers and Staff

- Who comes into the center? What are they coming for?
- What part of the city/neighborhoods people are coming from?
- What discussions, if any, have you had with residents?
- Are you typically approached? Or do you start the conversation?
- Do you know who comes in regularly? Are they homeowners or renters?
- What topics do you discuss with regulars?
- Do you talk with nonregulars too? What do you talk about?

- What did you learn about their experience?
- What information have you shared in the past few days?
- Who are you likely to share this information with? Why?
- Have you been able to serve everyone who comes in for help?
- What type of help do you provide?
- Do people request services or resources you cannot provide?
- Were they referred elsewhere?
- What do people need to do to receive help?
- Do different people want different kinds of help?
- What demographic groups are most represented from those coming in?
- Which ones seem most/least comfortable?

Community and Church Leaders and Field and Site Managers
- What are the events and circumstances that led to the establishment of this disaster relief/recovery center here?
- Who are the important actors and decision makers in the process?
- What changes in general coordination or service have you had to do? Why?

APPENDIX B

Reflections

My Positionality in the Field

As a researcher conducting fieldwork in New York City, I situated myself as a returning New Yorker to volunteer and research the unfortunate devastation that Superstorm Sandy had visited on fellow New Yorkers. My identity as a student was at the forefront of my interactions because introducing myself as such was part of my research consent process. I was warmly received by disaster survivors and disaster responders—community leaders, disaster field site managers, staff, and volunteers who were part of the disaster response efforts. Only four disaster survivors I approached for an interview refused. Most people I approached wanted to help me accomplish my research because of my student status. Some were impressed, or pleasantly surprised that I was going to earn a PhD. Interviews would end with comments like "I hope this helps you get your degree!" I told participants that primary motivation for pursuing this project was that learning of their experiences would help me convey to organizations responding to disasters how to better respond to future disasters. This continues to be my hope.

My positionality was often situational as I straddled the lines of race, class, gender, and citizenship. Mostly, I wanted to relate and be relatable to those with whom I would speak. I am a Black woman. I am an immigrant. I am a mother. I am also aware that certain aspects of my identity gained more salience in some interactions and contexts than other aspects of my identity. In Brooklyn, I fit right into Canarsie's Caribbean immigrant community, while in the Westville area in The Rockaways, in Queens, my Blackness stood out in stark contrast to the Whiteness of the survivors. To immigrants in both contexts, whether Black, Latinx, or White, I emphasized my St. Lucian background; I am from a small island in the Caribbean. In speaking to economically-privileged survivors, my affiliation with a private, "elite" university conjured familiarity in people with whom I spoke. In speaking with women who have children, my own motherhood conveyed unspoken angst that most mothers feel for their children's well-being. These points of mutual identification, or recognition, provided room in conversations for small talk and more openness.

Treatment of Racial Prejudice as Data

Despite these flexible positionalities, I could not negotiate my way out of my phenotype and the various shades of meanings projected onto my body by those who saw me and talked with me. Several studies have pointed to the importance of matching the race of researchers to that of respondents in order to elicit valid responses and produce valid analyses on race (Weinreb 2006; Stanfield 1993). Therefore, I did not solicit responses on an individual's racial attitudes. Requests to explicitly express racial sentiments through language were both unsolicited and unexpected.

Since this project partly relied on observations related to ascribed race, my minoritized racial status in White spaces necessarily meant that language or actions revealing racial prejudice could only be a conservative measure of interpersonal racism. Therefore, my discussions of actions that implicitly suggest racial prejudice are few and are relegated to inadvertent slips by interviewees. Alternatively, the bulk of my analysis centers on structural conceptualizations of ecologies of organizations and social networks and the structured relational experiences for racially minoritized and White disaster survivors and responders who participated in my research.

There were a few moments that revealed interpersonal prejudices. These were usually indirect accounts of disaster survivors speaking about someone else. Only one White male interviewee said, "Roy, he's my friend, but he is a racist. He hates Black people." Some of my race data was based on how participants interacted with me. For example, after a lengthy and quite pleasant interview with Peter, he smiled at me, nodding his head with accomplishment, saying, "See, see, I talk to Black people. I bet you thought I wasn't going to talk to you." As a Black woman living in the United States, I am aware of how cross-race interactions can work to reinforce ascriptions of subordinate and superordinate racial positions. Admittedly, despite this awareness, I refuse to cease being utterly perplexed by these situations. My internal monologue said, "Did he really just say that to me?" Then I at once reminded myself that my role there was that of a researcher. I courteously replied, "No not at all, the thought never even crossed my mind. Thanks for talking with me."

This experience made me more cognizant of the fact that my Blackness contextualized my interactions with participants, regardless of the genuine pleasantness of the conversations. For the rest of the day, I scrutinized all of my interactions. Did I fall into the color blindness trap? Did I really think that these Westvillers were blind to my race? Even as I was empathizing and sympathizing with them, were they simultaneously "othering" me? I decided to table these questions because I would never know whether, where, and when this actually occurred. This is why there is a notable absence of discussions of "racial prejudice." However, this silence speaks to the fact of my racial minoritization and the related fact that I could not be a suitable instrument for capturing this kind of data and not evidence of its absence in reality.

Researching in Abject Circumstances

Researchers trying to understand the plight of human beings living in abject circumstances have to be mindful that they may inadvertently further contribute to their trauma. I kept this awareness at the forefront of my mind at all times. If anyone hesitated to talk, even for a second, I instantly stopped describing my research. However, I would still feel conflicted that people's cheerful and welcoming gestures to speak with me may have been because they were lonely and would talk to anyone regardless of the purpose. These disaster survivors had lost everything, and here I was, asking them about their experiences because I really wanted to know on a personal level. However, the fact that this exchange would also result in a dissertation, articles, or a book made me feel like I was also benefiting from their misfortune. These feelings came and went. Other times, I felt fulfilled realizing that many of these disaster survivors would not have anyone with whom to process these feelings of helplessness, isolation, indignation, and despair at that moment when they needed it the most. I began to realize that I had more access to some survivors' perspectives and connected with them more closely than disaster responders at the FEMA- and NGO-managed response centers.

Researcher Interventions

As a field researcher, I was always aware of the fact that my very presence altered the dynamics of behavior in the field. Ethnographers are aware that we cocreate the reality that we are studying, but I still wished to minimize my influence on the setting and trajectory of events so that I preserved the opportunity to better understand the processes beyond outcomes. This was something I struggled with in the field. I began conducting my research in the early aftermath of the disaster. People's lives were in shambles. The inequalities across neighborhoods were becoming more evident in terms of survivors' awareness of the availability of resources that others had known about for weeks or even months.

I would soon realize that my back-and-forth movements across contexts and my gaining multiple viewpoints on the disaster made me an expert on the availability of disaster resources. Put simply, I had information that disaster survivors needed but did not know how to access or, worse yet, did even know that they needed to access. It was often heartbreaking to speak to disaster survivors from a high-poverty area or economically deprived survivors living in an economically privileged area, and while talking with them, I realized that they had no clue about certain services or assistance or had no local knowledge that others from more affluent areas had known of, applied for, and even received several weeks earlier. Understanding the processes that created and maintained such disparate speed in the diffusion of information across urban areas that housed different classes of disaster survivors became important to my continued inquiries.

After interviews, I often found myself providing helpful information to disaster survivors through the "cross-pollination" of what I learned from my visits to other disaster response centers, my conversations with disaster survivors and responders, and observations and meetings held in other neighborhoods. When I intervened, this led to a deeper understanding of nuances in experiences that I may have otherwise missed. I sought to benefit disaster survivors collectively, in real time, by sharing preliminary insights with responders. My approach to conversational interviewing also provoked field managers to think more deeply about issues of equity that were invisible to them before our conversation. My hope was that their actions could improve the lives of disaster survivors interfacing with these centers in real time. I amplified community voices by organizing a panel where founders of various local community-based organizations shared community perspectives and needs with an academic audience. However, more often than not, I found myself leaving the field frustrated that the problems that disaster survivors were experiencing were far too large and systemic for me to have any real impact other than through publishing and presenting my research, quite different from the public sociology I envisioned.

APPENDIX C

120 respondents in Brooklyn and The Rockaways

Study Participants	Study Population
AGE:	
18–29	16
30–39	18
40–49	24
50–59	31
60–69	21
70–79	10
GENDER:	
Male	56
Female	64
RACIAL OR ETHNIC GROUP	
Non-White	79
Other	13
SOCIOECONOMIC CLASS:	
Working class	24
Impoverished nonworking	33
Middle or upper class	63
FEMA OR STATE RESPONDERS:	
Field site managers	3
Volunteers or staff	6
NGO RESPONDERS:	
Field managers	2
Volunteers or staff	13
COMMUNITY-BASED RESPONDERS OR CHURCH RESPONDERS:	
Leaders, managers, or founders	10
Volunteers or staff	3
Community volunteers	10

REFERENCES

Adeola, Francis O., and J. Steven Picou. 2012. "Race, Social Capital, and the Health Impacts of Katrina: Evidence from the Louisiana and Mississippi Gulf Coast." *Human Ecology Review* 19(1): 10–24.

Aldrich, Daniel P. 2012. *Building Resilience: Social Capital in Post-Disaster Recovery*. Chicago: University of Chicago Press.

Baradaran, Mehrsa. 2017. *The Color of Money: Black Banks and the Racial Wealth Gap*. Cambridge, Mass.: Harvard University Press.

Barnshaw, John. 2005. "The Continuing Significance of Race and Class among Houston Hurricane Katrina Evacuees. *Natural Hazards Observer* 30(2): 11–13.

Barnshaw, John, and Joseph Trainor. 2010. "Race, Class and Capital amidst the Hurricane Katrina Diaspora." In *The Sociology of Katrina: Perspectives on a Modern Catastrophe*, 2nd ed, edited by D. Brunsma, D. Overfelt, and S. Picou, 103–18. Boston: Rowman & Littlefield.

Barth, Fredrik. 1969. *Ethnic Groups and Boundaries: The Social Organization of Culture Difference*. Oslo: Universitetsforlaget.

Bates, Frederick. 1982. *Recovery, Change, and Development*. Athens: University of Georgia Press.

Bellot, Alfred H. 1917. *History of the Rockaways: 1685 to 1917*. Far Rockaway, N.Y.: Bellot's Histories.

Blumer, Herbert. 1958. "Race Prejudice as a Sense of Group Position." *Pacific Sociological Review* 1(1): 3–7.

Bobo, Lawrence, and Vincent L. Hutchings. 1996. "Perceptions of Racial Group Competition: Extending Blumer's Theory of Group Position to a Multiracial Social Context." *American Sociological Review* 61: 951–72.

Bolger, Daniel. 2021. "The Racial Politics of Place in Faith-Based Social Service Provision." *Social Problems* 68(3): 535–51.

Bolin, Bob. 2006. "Race, Class, Ethnicity, and Disaster Vulnerability." In *Handbook of Disaster Research*, edited by Havidán Rodriguez, Enrico Quarantelli, and Russell Dynes, 107–31. Memphis: Central United States Earthquake Consortium, Monograph no. 5.

Bonilla-Silva, Eduardo. 1997. "Rethinking Racism: Toward a Structural Interpretation." *American Sociological Review* 62(3): 465–80.

Bourdieu, Pierre. 1986. "The Forms of Capital." In *Handbook of Theory and Research for the Sociology of Education*, edited by J. G. Richardson, 242. Westport, Conn.: Greenwood Press.

Bourdieu, Pierre, and Jean Claude Passeron. 1977. *Reproduction in Education, Society and Culture*. Thousand Oaks, Calif.: SAGE.

Breiger, Ronald L. 1974. "The Duality of Persons and Groups." *Social Forces* 53(2): 181–90.

Brooklyn Public Library. 2016. "Black Canarsie: A History." February 1, 2016. https://www.bklynlibrary.org/locations/jamaica-bay/black-canarsie-history.

Brunsma, David L., David Overfelt, and J. Steven Picou, eds. 2007. *The Sociology of Katrina: Perspectives on a Modern Catastrophe*. Lanham, Md.: Rowman & Littlefield.

Bullard, Robert D. 1983. "Solid Waste Sites and the Black Houston Community." *Sociological Inquiry* 53(2–3): 273–88.

———. 2009. *Race, Place and Environmental Justice after Hurricane Katrina: Struggles to Reclaim, Rebuild, and Revitalize New Orleans and the Gulf Coast*. Boulder, Colo.: Westview Press.

Bullard, Robert D., and Beverly Wright. 2012. *The Wrong Complexion for Protection: How the Government Response to Disaster Endangers African American Communities*. New York: New York University Press.

Burawoy, Michael. 2003. "Revisits: An Outline of a Theory of Reflexive Ethnography." *American Sociological Review* 68(5): 645–79.

Calarco, Jessica McCrory. 2011. "'I Need Help!' Social Class and Children's Help-Seeking in Elementary School." *American Sociological Review* 76(6): 862–82.

Caro, Robert A. 2006. *The Power Broker: Robert Moses and the Fall of New York*. Francis Parkman Prize ed. History Book Club. Available online from EBSCO.

Charles, Camille Zubrinsky. 2003. "The Dynamics of Racial Residential Segregation." *Annual Review of Sociology* 29: 167–207.

Coleman, James S. 1990. *Foundations of Social Theory*. Cambridge, Mass.: Belknap Press of Harvard University Press.

Collard, Rosemary-Claire, and Jessica Dempsey. 2017. "Capitalist Natures in Five Orientations." *Capitalism Nature Socialism* 28(1): 78–97.

Cutter, Susan, Christopher T. Emrich, Jerry T. Mitchell, Walter W. Piegorsch, Mark M. Smith, and Lynn Weber. 2014. *The Recovery Divide: Hurricane Katrina and the Forgotten Coast of Mississippi*. Cambridge: Cambridge University Press.

Cutter, Susan L., Jerry T. Mitchell, and Michael S. Scott. 2000. "Revealing the Vulnerability of People and Places: A Case Study of Georgetown County, South Carolina." *Annals of the Association of American Geographers* 90(4): 713–37.

David, Emmanuel, and Elaine Enarson, eds. 2012. *The Women of Katrina: How Gender, Race, and Class Matter in American Disaster*. Nashville: Vanderbilt University Press.

Desmond, Matthew. 2012. "Disposable Ties and the Urban Poor." *American Journal of Sociology* 117(5): 1295–335.

———. 2016. *Evicted: Poverty and Profit in the American City*. New York: Crown.

Desmond, Matthew, and Nathan Wilmers. 2019. "Do the Poor Pay More for Housing?

Exploitation, Profit, and Risk in Rental Markets." *American Journal of Sociology* 124(4): 1090–124.

Donner, William, and Havidan Rodriguez. 2008. "Population Composition, Migration and Inequality: The Influence of Demographic Changes on Disaster Risk and Vulnerability." *Social Forces* 87(2): 1089–114.

Drabek, Thomas E., and William H. Key. 1984. *Conquering Disaster: Family Recovery and Long-Term Consequences*. New York: Irvington.

Dynes, Russell R. 2002. "The Importance of Social Capital in Disaster Response." University of Delaware Disaster Research Center. https://udspace.udel.edu/bitstream/handle/19716/292/PP%20327.pdf?sequence=1&isAllowed=y.

Eggers, Frederick, and Fouad Moumen. 2010. *Investigating Very High Rent Burdens among Renters in the American Housing Survey*. Washington, D.C.: U.S. Department of Housing and Urban Development.

Elliott, James R., Timothy J. Haney, and Petrice Sams-Abiodun. 2010. "Limits to Social Capital: Comparing Network Assistance in Two New Orleans Neighborhoods Devastated by Hurricane Katrina." *Sociological Quarterly* 51(4): 624–48.

Erikson, Kai. 1976. *Everything in Its Path: Destruction of Community in the Buffalo Creek Flood*. New York: Simon & Schuster.

Faber, Jacob William. 2015. "Superstorm Sandy and the Demographics of Flood Risk in New York City." *Human Ecology* 43(3): 363–78.

FEMA (Federal Emergency Management Agency). 1998. "Overview of Stafford Act Support to States." https://www.fema.gov/pdf/emergency/nrf/nrf-stafford.pdf.

———. 2017. "Remembering Sandy Five Years Later." October 28, 2017. https://www.fema.gov/press-release/20210318/remembering-sandy-five-years-later.

———. 2018. "FACT SHEET: FEMA Eligibility Letter May Not Be Last Word on Disaster Assistance." December 15, 2018. https://www.fema.gov/press-release/20210318/fact-sheet-fema-eligibility-letter-may-not-be-last-word-disaster-assistance.

———. 2021. "Fact Sheet: Don't Wait to Begin Cleaning, Making Repairs." March 18, 2021. https://www.fema.gov/press-release/20210318/fact-sheet-dont-wait-begin-cleaning-making-repairs.

Foner, Nancy. 2000. *From Ellis Island to JFK: New York's Two Great Waves of Immigration*. New Haven: Yale University Press.

———. 2005. *In a New Land: A Comparative View of Immigration*. New York: New York University Press.

Fothergill, Alice, and Lori A. Peek. 2004. "Poverty and Disasters in the United States: A Review of Recent Sociological Findings." *Natural Hazards* 32(1): 89–110.

Fraser, Nancy. 2016. "Expropriation and Exploitation in Racialized Capitalism: A Reply to Michael Dawson." *Critical Historical Studies* 3(1): 163–78.

Frazier, Edward Franklin. 1957. *Black Bourgeoisie*. New York: Macmillan.

Fritz, C. E. 1961. "Disasters." In *Contemporary Social Problems*, edited by Robert K. Merton and Robert A. Nisbet, 651–94. New York: Harcourt, Brace & World.

Fussell, Elizabeth. 2012. "Help from Family, Friends, and Strangers during Hurricane Katrina 2012." In *Displaced: Life in the Katrina Diaspora*, edited by Lynn Weber and Lori Peek, 150–66. Austin: University of Texas Press.

Galtung, Johan. 1969. Violence, Peace, and Peace Research. *Journal of Peace Research*, 6(3): 167–91.
GAO (Government Accountability Office). 2015. "FEMA Has Made Progress since Hurricanes Katrina and Sandy, but Challenges Remain." October 22, 2015. https://www.gao.gov/assets/gao-16-90t.pdf.
Gibbs, Linda, and Caswell F. Holloway. 2013. *Hurricane Sandy after Action: Report and Recommendations to Mayor Michael R. Bloomberg*. May 2013. https://www1.nyc.gov/assets/housingrecovery/downloads/pdf/2017/sandy_aar_5-2-13.pdf.
Glazer, Nathan, and Daniel P. Moynihan. 1970. *Beyond the Melting Pot: The Negroes, Puerto Ricans, Jews, Italians, and Irish of New York City*. 2nd ed. Cambridge, Mass.: MIT Press.
Gotham, Kevin Fox, and Krista Brumley. 2002. "Using Space: Agency and Identity in a Public-Housing Development." *City & Community* 1(3): 267–89.
Gotham, Kevin Fox, and Miriam Greenberg. 2014. *Crisis Cities: Disaster and Redevelopment in New York and New Orleans*. New York: Oxford University Press.
Granovetter, Mark S. 1973. "The Strength of Weak Ties." *American Journal of Sociology* 78(6): 1360–80.
———. 1985. "Economic Action and Social Structure: The Problem of Embeddedness." *American Journal of Sociology* 91(3): 481–510.
Gunaratnam, Yasmin. 2003. *Researching "Race" and Ethnicity: Methods, Knowledge, and Power*. Thousand Oaks, Calif.: SAGE.
Hansen, Karen V. 2011. "The Asking Rules of Reciprocity." In *At the Heart of Work and Family: Engaging the Ideas of Arlie Hochschild*, edited by Anita Ilta Garey and Karen V. Hansen, 112–23. New Brunswick, N.J.: Rutgers University Press.
Herd, Pamela, and Donald P. Moynihan. 2019. *Administrative Burden: Policymaking by Other Means*. New York: Russell Sage Foundation.
Hernández, Diana, David Chang, Carole Hutchinson, Evanah Hill, Amenda Almonte, Rachel Burns, Peggy Shepard, Ingrid Gonzalez, Nora Reissig, and David Evans. 2018. "Public Housing on the Periphery: Vulnerable Residents and Depleted Resilience Reserves Post-Hurricane Sandy." *Journal of Urban Health* 95(5): 703–15.
Heron, Melonie P. 2001. *The Occupational Attainment of Caribbean Immigrants in the United States, Canada, and England*. New York: LFB Scholarly.
Hoelscher, Steven. 2003. "Making Place, Making Race: Performances of Whiteness in the Jim Crow South." *Annals of the Association of American Geographers* 93(3): 657–86.
Hunter, Albert. 1974. *Symbolic Communities: The Persistence and Change of Chicago's Local Communities*. Chicago: University of Chicago Press.
INCITE! 2007. Women of Color Against Violence, eds. *The Revolution Will Not Be Funded: Beyond the Non-Profit Industrial Complex*. Cambridge, Mass.: South End Press.
———. 2020. *The Revolution Will Not Be Funded: Beyond the Non-Profit Industrial Complex*. Rev. ed. Durham, N.C.: Duke University Press.
Jung, Moon-Kie, and Yaejoon Kwon. 2020. "The Racial State in the Age of Racial Formation Theory and Beyond." In *The New Handbook of Political Sociology*, edited by Thomas Janoski, Cedric de Leon, Joya Misra, and Isaac William Martin, 1003–26. Cambridge: Cambridge University Press.

Kaplan, Lawrence, and Carol P. Kaplan. 2003. *Between Ocean and City: The Transformation of Rockaway, New York*. New York: Columbia University Press.

Kasinitz, Phillip. 2008. *Inheriting the City: The Children of Immigrants Come of Age*. New York: Russell Sage Foundation.

Kasinitz, Philip, John H. Mollenkopf, and Mary C. Waters. 2004. *Becoming New Yorkers: Ethnographies of the New Second Generation*. New York: Russell Sage Foundation.

Kent, Mary Mederios. 2007. "Immigration and America's Black Population." *Population Bulletin* 62(4).

King, Rita J. 2009. "Post-Katrina Profiteering: The New Big Easy." In *Race, Place, and Environmental Justice after Hurricane Katrina: Struggles to Reclaim, Rebuild, and Revitalize New Orleans and the Gulf Coast*, edited by Robert D. Bullard and Beverly Wright, 169–82. New York: Routledge.

Klein, Naomi. 2008. *The Shock Doctrine: The Rise of Disaster Capitalism*. New York: Henry Holt.

Kliff, Sarah. 2013. "The Irish-American Population Is Seven Times Larger than Ireland." *Washington Post*, March 17, 2013. https://www.washingtonpost.com/news/wonk/wp/2013/03/17/the-irish-american-population-is-seven-times-larger-than-ireland/.

Klinenberg, Eric. 2002. *Heat Wave: A Social Autopsy of Disaster in Chicago*. Chicago: University of Chicago Press, 2002.

———. 2004. "Overheated." *Contemporary Sociology* 33(5): 521–28.

———. 2015. *Heat Wave: A Social Autopsy of Disaster in Chicago*. Chicago: University of Chicago Press.

Krause, Neal. 2008. *Aging in the Church: How Social Relationships Affect Health*. West Conshohocken, Penn.: Templeton Foundation Press.

Lamont, Michèle, and Marcel Fournier. 1992. *Cultivating Differences: Symbolic Boundaries and the Making of Inequality*. Chicago: University of Chicago Press.

LaRossa, Ralph. 2005. "Grounded Theory Methods and Qualitative Family Research." *Journal of Marriage and Family* 67(4): 837–57.

Laster Pirtle, Whitney N. 2020. "Racial Capitalism: A Fundamental Cause of Novel Coronavirus (Covid-19) Pandemic Inequities in the United States." *Health Education & Behavior* 47(4): 504–8.

Lawler, Edward J., Shane R. Thye, and Jeongkoo Yoon. 2009. *Social Commitments in a Depersonalized World*. New York: Russell Sage Foundation.

Lieberson, Stanley. 2000. *Matter of Taste: How Names, Fashions, and Culture Change*. New Haven: Yale University Press.

Lin, Nan. 2000. "Inequality in Social Capital." *Contemporary Sociology* 29(6): 785–95.

———. 2001. *Social Capital: A Theory of Social Structure and Action*. New York: Cambridge University Press.

Lin, Nan, Karen S. Cook, and Ronald S. Burt, eds. 2001. *Social Capital: Theory and Research*. New York: Aldine de Gruyter.

Lin, Nan, Walter M. Ensel, and John C. Vaughn. 1981. "Social Resources and Strength of Ties: Structural Factors in Occupational Status Attainment." *American Sociological Review* 46(4): 393–405.

Lin, Nan, Yang-Chih Fu, and Ray-May Hsung. 2001. "The Position Generator: Measurement Techniques for Investigation of Social Capital." In *Social Capital: Theory and Research*, edited by Nan Lin, Karen S. Cook, and Ronald S. Burt, 57–81. New York: Aldine de Gruyter.

Lin, Nan, Mary W. Woelfel, and Stephen C. Light. 1985. "The Buffering Effect of Social Support Subsequent to an Important Life Event." *Journal of Health and Social Behavior* 26(3): 247–63.

Litt, Jacquelyn. 2012. "'We Need to Get Together with Each Other': Women's Narratives of Help in Katrina's Displacement." In *Displaced: Life in the Katrina Diaspora*, edited by Lynn Weber and Lori Peek, 167–82. Austin: University of Texas Press.

Litt, Jacquelyn, Althea Skinner, and Kelley Robinson. 2012. "The Katrina Difference: African American Women's Networks and Poverty in New Orleans after Katrina." In David and Enarson, *Women of Katrina*, 130–41.

Luft, Rachel E. 2016. "Disaster Patriarchy: An Intersectional Model for Understanding Disaster at the Ten-Year Anniversary of Hurricane Katrina." *Feminist Formations* 28(2): 1–26.

Marwell, Nicole P. 2007. *Bargaining for Brooklyn: Community Organizations in the Entrepreneurial City*. Chicago: University of Chicago Press.

Massey, Douglas S., and Nancy A. Denton. 1993. *American Apartheid: Segregation and the Making of the Underclass*. Cambridge, Mass.: Harvard University Press.

Mayorga, Sarah. 2014. *Behind the White Picket Fence: Power and Privilege in a Multiethnic Neighborhood*. Chapel Hill: University of North Carolina Press.

Model, Suzanne. 2008. *West Indian Immigrants: A Black Success Story?* New York: Russell Sage Foundation.

Mooney, Margarita A. 2009. *Faith Makes Us Live: Surviving and Thriving in the Haitian Diaspora*. Berkeley: University of California Press.

NAC (National Advisory Council). 2020. "National Advisory Council Report to the FEMA Administrator." November 2020. https://www.fema.gov/sites/default/files/documents/fema_nac-report_11-2020.pdf.

O'Leary, Zina. 2005. *Researching Real-World Problems: A Guide to Methods of Inquiry*. London: SAGE.

Oliver, Melvin L., and Thomas M. Shapiro. 1995. *Black Wealth/White Wealth: A New Perspective on Racial Inequality*. New York: Routledge.

Oliver-Smith, Anthony. 1986. *The Martyred City: Death and Rebirth in the Andes*. Albuquerque: University of New Mexico Press.

Omi, Michael, and Howard Winant. 1986. *Racial Formation in the United States: From the 1960s to the 1990s*. New York: Routledge.

———. 1994. *Racial Formation in the United States: From the 1960s to the 1990s*. Rev. ed. New York: Routledge.

Passel, Jeffery, and Rebecca L. Clark. 1999. "Immigrants in New York: Their Legal Status, Incomes, and Taxes." Urban Institute. http://webarchive.urban.org/publications/407432.html.

Passel, Jeffrey S., and D'Vera Cohn. 2011. "Unauthorized Immigrant Population: National and State Trends, 2010." Pew Hispanic Center. https://www.pewresearch.org/hispanic/2011/02/01/unauthorized-immigrant-population-brnational-and-state-trends-2010/.

Patterson, Orlando. 1982. *Slavery and Social Death: A Comparative Study*. Cambridge, Mass.: Harvard University Press.

Peacock, Walter Gillis, Hugh Gladwin, and Betty Hearn Morrow, eds. 1997. *Hurricane Andrew and the Reshaping of Miami: Ethnicity, Gender and the Sociology of Disasters*. Gainesville: University Press of Florida.

Pelling, Michael. 2003. *Natural Disaster and Development in a Globalizing World*. New York: Routledge.

Pew Hispanic Center. 2006. "Modes of Entry for the Unauthorized Migrant Population." May 22, 2006. https://www.pewresearch.org/hispanic/2006/05/22/modes-of-entry-for-the-unauthorized-migrant-population/.

Picou, J. Steven, Brent K. Marshall, and Duane A. Gill. 2004. "Disaster, Litigation, and the Corrosive Community." *Social Forces* 82(4): 1493–522.

Portes, Alejandro. 1998. "Social Capital: Its Origins and Applications in Modern Sociology." *Annual Review of Sociology* 24(1):1–24.

Portes, Alejandro, and Rubén G. Rumbaut. 2006. *Immigrant America: A Portrait*. Berkeley: University of California Press.

Portes, Alejandro, and Julia Sensenbrenner. 1993. "Embeddedness and Immigration: Notes on the Social Determinants of Economic Action." *American Journal of Sociology* 98(6): 1320.

Pulido, Laura. 2000. "Rethinking Environmental Racism: White Privilege and Urban Development in Southern California." *Annals of the Association of American Geographers* 90(1): 12–40.

———. 2016. "Flint, Environmental Racism, and Racial Capitalism." *Capitalism Nature Socialism* 27(3): 1–16.

Ray, Victor. 2019. "A Theory of Racialized Organizations." *American Sociological Review* 84(1): 26–53.

Reid, Megan. 2012. "Mothering after a Disaster: The Experiences of Black Single Mothers Displaced by Hurricane Katrina." In David and Enarson, *Women of Katrina*, 105–17.

Rieder, Jonathan. 1985. *Canarsie: The Jews and Italians of Brooklyn against Liberalism*. Cambridge, Mass.: Harvard University Press.

Ritchie, Liesel Ashley. 2004. "Voices of Cordova: Social Capital in the Wake of the Exxon Valdez Oil Spill." PhD diss., Mississippi State University.

Robinson, Cedric. 1983. *Black Marxism: The Making of the Black Radical Tradition*. Chapel Hill: University of North Carolina Press.

———. 2000. *Black Marxism. The Making of the Black Radical Tradition*. 2nd ed. Chapel Hill: University of North Carolina Press.

Ryan, Dennis P. 1999. *A Journey through Boston Irish History*. Images of America: Massachusetts. Mount Pleasant, S.C.: Arcadia.

Scott, Janny. 2001. "Amid New York's Sea of Faces, Islands of Segregation." *New York Times*, June 18, 2001. https://www.nytimes.com/2001/06/18/nyregion/amid-new-yorks-sea-of-faces-islands-of-segregation.html.

Sidanius, Jim, and Felicia Pratto. 1999. *Social Dominance: An Intergroup Theory of Social Hierarchy and Oppression*. Cambridge: Cambridge University Press.

Small, Mario Luis. 2009a. "'How Many Cases Do I Need?': On Science and the Logic of Case Selection in Field Research." *Ethnography* 10: 5–38.

———. 2009b. *Unanticipated Gains: Origins of Network Inequality in Everyday Life.* New York: Oxford University Press.

Smith, Sandra Susan. 2005. "'Don't Put My Name on It': Social Capital Activation and Job Finding Assistance among the Black Urban Poor." *American Journal of Sociology* 111(1): 1–57.

Solnit, Rebecca. 2010. *A Paradise Built in Hell: The Extraordinary Communities That Arise in Disaster.* New York: Penguin Books.

Stack, Carol B. 1974. *All Our Kin: Strategies for Survival in a Black Community.* New York: Harper & Row.

Stanfield, John H., II. 1993. "Epistemological Considerations." In *Race and Ethnicity in Research Methods*, edited by John H. Stanfield II and Routledge M. Dennis. Newbury Park, Calif.: SAGE.

Taylor, Dorceta E. 2014. *Toxic Communities: Environmental Racism, Industrial Pollution, and Residential Mobility.* New York: New York University Press.

Tilly, Charles. 1999. *Durable Inequality.* Berkeley, Calif.: University of California Press.

Tomaskovic-Devey, Donald, and Dustin Robert Avent-Holt. 2019. *Relational Inequalities: An Organizational Approach.* New York: Oxford University Press.

United Church of Christ, Commission for Racial Justice. 1987. "Toxic Wastes and Race in the United States: A National Report on the Racial and Socio-Economic Characteristics of Communities with Hazardous Waste Sites." https://www.nrc.gov/docs/ML1310/ML13109A339.pdf.

U.S. Congress. 2002. "H.R. 5005—107th Congress (2001–2002): Homeland Security Act of 2002." November 19, 2002. https://www.congress.gov/bill/107th-congress/house-bill/5005.

———. 2006. "S.3721—109th Congress (2005–2006): Post-Katrina Emergency Management Reform Act of 2006." August 3, 2006. https://www.congress.gov/bill/109th-congress/senate-bill/3721.

Vargas, Robert. 2019. "Gangstering Grants: Bringing Power to Collective Efficacy Theory." *City and Community* 18(1): 369–91.

Waters, Mary C. 1999. *Black Identities: West Indian Immigrant Dreams and American Realities.* Cambridge, Mass.: Harvard University Press.

Warren, Robert. 2003. "Estimates of the Unauthorized Immigrant Population Residing in the United States: 1990 to 2000." Washington, D.C.: U.S. Immigration and Naturalization Service, Office of Policy and Planning.

Weber, Lynn. 2017. "Through the Fog of Disaster: The Process of (Re)Creating Inequities." *Research Counts*, November 29. Natural Hazards Center, University of Colorado, Boulder. https://hazards.colorado.edu/news/research-counts/through-the-fog-of-disaster-the-process-of-re-creating-inequities.

Weber, Lynn, and Lori Peek, eds. 2012. *Displaced: Life in the Katrina Diaspora.* Austin: University of Texas Press.

Weinreb, Alexander A. 2006. "The Limitations of Stranger-Interviewers in Rural Kenya." *American Sociological Review* 71(6): 1014–39.

Wilson, William Julius. 2012. *The Truly Disadvantaged: The Inner City, the Underclass, and Public Policy.* Chicago: University of Chicago Press.

Woolcock, Michael. 1998. "Social Capital and Economic Development: Toward a Theoretical Synthesis and Policy Framework." *Theory and Society* 27(2): 151–208.

Woolcock, Michael, and Deepa Narayan. 2000. "Social Capital: Implications for Development Theory, Research, and Policy." *World Bank Research Observer* 15(2): 225–49.

Wooten, Melissa E. 2015. *In the Face of Inequality: How Black Colleges Adapt*. Albany: State University of New York Press.

Wooten, Melissa, and Andrew J. Hoffman. 2016. "Organizational Fields Past, Present and Future." In *The SAGE Handbook of Organizational Institutionalism*, 2nd ed., edited by R. Greenwood, C. Oliver, K. Sahlin, and R. Suddaby, 55–74. London: SAGE.

Data Sources for Figures

Illustrator: Whitley Plummer

Figs. 1 & 3. NYC OpenData and 2010 Census Cartographic Boundary Files

Fig. 4. CDC 2010 Social Vulnerability; NYC Open Data, NYC Area Tabulations, Roadbed shapefile. CDC Description of Social Vulnerability Index: "SVI indicates the relative vulnerability of every U.S. Census tract. Census tracts are subdivisions of counties for which the Census collects statistical data. SVI ranks the tracts on 15 social factors, including unemployment, minority status, and disability, and further groups them into four related themes. Thus, each tract receives a ranking for each Census variable and for each of the four themes, as well as an overall ranking" (CDC SVI 2018 documentation, p.1)

Figs. 5 & 9. 2012 ACS 5yr. Using Ancestry data from the 2012 ACS 5 yr, data was pulled from census.gov to identify the breakdown of ethnic groups in the region. Some limitations are: Many people may not have identified explicitly with an ancestral background or ethnic group, even though they might be a descendant of one of these classifications. A significant portion of the population identified as West Indian (with no further breakdown). Additionally, some respondents identified as American; this provides no clear understanding of what their ethnic grouping/designation might be as American represents/encompasses many different groups.

Fig. 7. CDC 2010 Social Vulnerability; NYC Open Data—NYC Area Tabulations, Roadbed shapefile

Fig. 8. NYC OpenData, 2010 Census Cartographic Boundary Files

Fig. 10. (1) ACS 2012 5 yr Foreign Born; (2) NYC OpenData (Created October 26, 2012)

Fig. 11. The FIRM maps reflect the designations FEMA and NYC made in their own maps; the classifications were condensed into three groups: A, V, and X shaded.
    A High Risk, 1% chance of flood.
    V High Risk, 1% chance of flood but is in a coastal region, so wave action is likely.
    X Low Risk, a .2% chance of flood.

Fig. 12. NYC OpenData. Evacuation Zones: The Evacuations are designations made by NYC. NYC's description of the evacuation zones is as follows: Hurricane evacuation zones are used for communicating evacuation orders to the public. These zones indicate the areas at most risk of flooding due to storm surge during a hurricane. Zone A is the most at risk, followed by zone B, and then zone C.

INDEX

Page numbers in italics indicate illustrations; those with a *t* indicate tables.

African Americans. *See* Black Americans
alcoholism, 96
altruism: local, 73, 115–17, 118–20; Solnit on, 76–77. *See also* volunteerism
Always With You (org.), 7, 23, 26–28, 80, 145; church aid groups and, 116; location of, 28–29; Resiliency Is Us and, 28–30, 74–75, 121, 128–29; volunteers of, 73–74
Apple Angels (org.), 73–74, 77
asking rules, 70
Avent-Holt, Dustin Robert, 19

basement renters, 37, 134, 138; documentation of, 43–44; FEMA assistance to, 37–40, 81; housing assistance for, 40–45; multigenerational living arrangements of, 37–38. *See also* landlords
Black Americans, 10, 69; in Canarsie, 33–36, *35*, 35t, 117–18; enslavement of, xv–xvii; Great Migration of, 16, 24, 33; Jews and, 27; racial capitalism and, xv–xx; in Rockaways, 24–26, *25*, 26t. *See also* race/class influences
boarding houses, 28
Bolger, Daniel, 20
Bonilla-Silva, Eduardo, 16
bootstrap bias, 85
Brooklyn, 111, 115–17, 124–27, 144–46

Canarsie, 51; church aid groups in, 2, 121–23; community-based organizations in, 36; demographics of, 30–36, *34*, 35t; disaster response in, 40–42, 117–18, 122; FEMA in, 2, 40–42, 117–18, 121–23; flood zone designations of, 48, *49*; maps of, *4*, *23*, *33*; organization hosting of, 121–23; Resiliency Is Us in, 118, 122; social vulnerability index of, *32*
Caribbean immigrants, 9; in Canarsie, 32–36, *34*, 54–55, 117; in Rockaways, 25, 111
Chicago heatwave (1995), 16–17
Chicago school economists, xvii
church aid groups, 70, 75–78, 114–29, 143–45; coalitions of, 144–46; community-based organizations and, xix–xx, 19; undocumented immigrants and, 32–33
Clarke, Yvette, 35
class, 5, 131; bootstrap bias and, 78, 84–85; segmentation by, 34–36. *See also* race/class influences
color blindness, 86–88, 153
Colored Colony, 34
community-based organizations, xix–xx, 36; crisis capital of, 71–73, 76–77; networking of, 51, 113–29; racializing of, 18–20. *See also* volunteerism
Covid-19 pandemic, 7, 135, 144
criminal justice system, xvii, 112, 128
crisis capital, 6, 117, 129; of community-based organizations, 72–74, 75–78; definition of, 67; Resiliency Is Us and, 76–78. *See also* social capital
crisis hotline, 82–83
Cutter, Susan L., xxii–xxiii

Department of Homeland Security (DHS), 48
Desmond, Matthew, xvii
disaster response, 41–43, 81–82, 101–4, 114, 127; logic of, 6, 78–83, 87, 98; by New York State, 6–7, 78, 81–83, 87–88, 131; race/class influences in, 26–29, 109, 114–29, 131–32; racializing processes in, *17*, 17–19; types of, 7, 113–29, 144–46
drug rehabilitation programs, 97; homelessness and, 68–69, 92–99; poverty and, 91–92. *See also* mental health services

Eastville, 52, 100, 120; Always With You in, 26–28, 73–75, 117, 128–29; information diffusion in, 84, 111–12; Resiliency Is Us in, 76–78, 107–8, 129–30; social capital loss in, 67–70. *See also* Rockaways
ecology: of disadvantage, 5, 91–99; of disaster response organizations, 14, 115–30; of inequity, 4–5, 15–21, *17*, 29, 131–32; of privilege, 5, 7, 99
El Salvadoran immigrants, 26–27
ethnographic analysis, 13–14
ethnographic discovery, 1, 12
exclusionary narratives, 108–13

Federal Emergency Management Agency (FEMA), 48, 91–98; church aid groups and, 118–21, 143–44; coverage guidelines of, 37–38; Department of Homeland Security and, 48; Disaster Response Center of, 40–42; follow-up mechanisms of, 52, 56–61; helpline of, 46, 50, 52, 54, 81; independent contractors of, 64–65; after Katrina, 48, 56; Small Business Administration and, 56–62; transitional housing of, 92; Voluntary Agency Liaison of, 126
flood insurance, 40, 44, 56–58, 62–63, 143
food insecurity, 53–54
Fraser, Nancy, xvi, xvii
Friedman, Milton, xvii

Galtung, Johan, xvi–xvii, 16
gender, xvi; informal economy and, 32
Glazer, Nathan, 32
Gotham, Kevin Fox, xxii
Great Migration (1910–1950), 17, 26, 34
Greenberg, Miriam, xxii
Greenpeace International, 145
gun violence, 144
Gunaratnam, Yasmin, xv

Haitian immigrants, 84–85; in Canarsie, 31–34, *34*, 40; Mooney on, 32–33; in Rockaways, 112; SBA loans to, 60–61
Hart-Celler immigration law (1965), 33
Hilliard, Raymond M., 26
Hoffman, Andrew J., 20
Homeland Security Act (2002), 48
homelessness, 45; drug rehabilitation programs and, 67–68, 91–98; elderly and, 81; relocation program for, 83; shelters for, 45
homeowners, 37, 79, 88–90; basement renters versus, 39–41, 44; FEMA assistance to, 56, 59, 60, 63, 65, 144; social capital loss of, 70–72
housing assistance, xvii, 24, 41–46, 89; FEMA grants for, 56, 59, 60, 63, 65; gender and, 25, 92–93; real estate market and, 25–26; social services economy and, xviii–xix; transitional, 68–69, 91–98. *See also* public housing
Housing and Urban Development (HUD) Department, 123

immigrants, 25–26, 85; documented, 36, 40; undocumented, xvii, 31–32, 42, 45–47, 112
inclusionary projects, 108
Individuals and Households Program (IHP), 56
inequity, 50; definition of, 16; Tilly on, 22
information diffusion, 84, 101, 106–7, 111–12
insurance coverage, 40, 44, 56–58, 61–62, 143
interviews, 147–50; demographics of, 8t, 9–10; self-positionality of, 151–55
Irish immigrants, 26, 35, 101, 109
Italian immigrants, 34–35

Jamaican immigrants, 34–35
Jews, 26, 34–35
Jung, Moon-Kie, xvi

Katrina (2005), 20–21; FEMA reforms after, 48, 56; Sandy and, xx–xxiii
King, Rita J., xxii
Klein, Naomi, xxii
Kliff, Sarah, 25
Klinenberg, Eric, 13–16
Kwon, Yaejoon, xvi

landlords, 39–41, 43; discrimination against large families by, 45; profits of, xvii–xix; Rapid Repairs program for, 43–44, 88, 121; rent to value ratio and, xvii–xviii
LaRossa, Ralph, 13

Index

Latinos, 20, 74, 111; racial capitalism and, xvii–xx; in Rockaways, 25–27, 26t, 27
Lin, Nan, 18
logic of disaster response, 6, 78–83, 87, 98
Long-Term Recovery (org.), 78, 80, 124

mental health services, 26, 80–82, 110. *See also* drug rehabilitation programs
methodology, 1–3, 12–13; interview questions of, 148–51; limitations of, 154–55; nested processes/structures of, 5, 15, 17, 21; self-positionality of, 151–53
Mexican immigrants, 111. *See also* Latinos
middle-class bootstrap bias, 134
Mooney, Margarita A., 34
Moses, Robert, 25
Moynihan, Patrick, 26

Native Americans, xvi, 10, 25, 75
neoliberalism, xvii
networking, 5–6, 143–44; information diffusion and, 84, 106–7; organizational, 51, 113–29; race/class influences in, 14–21; as social capital, 20–21; women-centered, 19–20. *See also* social capital
New York Disaster Interfaith Services (NYDIS), 126
New York State (NYS) disaster response centers, 6–7, 78, 81–83, 87–88, 131

Occupy Sandy (org.), 145
Omi, Michael, 16
organization agglomeration, 7, 113–21, 128
organization coalition, 7, 113, 115, 124–27, 144–46
organization hosting, 7, 114, 121–23
organization isolation, 7, 114, 129–30

Patterson, Orlando, xvi
Peninsula Circles (org.), 121
Polish immigrants, *34*
Portes, Alejandro, 85
Post-Katrina Emergency Management Reform Act (2006), 47, 55
prison system, xvii, 112, 128
public housing, xvii, 24, 26, 88; "urban renewal" policies and, 34; vouchers for, 46. *See also* housing assistance
Puerto Ricans, xvi, 27–28, 35
Pulido, Laura, 16–17

race/class influences, xv–xx, 4–5, *17*, 23–30, *25*; and color blindness, 39, 78, 85–87, 153; on disaster response, 26–29, 109, 114–29, 131–32; on networks, 15–22; on research, 151–53; of social vulnerability, 15, 33, 54, 88; on volunteerism, 28
racial capitalism, xv–xx
racializing organizations, 18–20
Rapid Repairs program, 43–44, 87, 120
Ray, Victor, 19
Reagan, Ronald, xvii
Red Hook, 51
Reid, Megan, 52, 54
relational capital, 6
rent to value ratio, xvii–xviii. *See also* landlords
researcher interventions, 1, 153–54. *See also* methodology
Resiliency Is Us (org.), 23, 28, 107–8; Always With You and, 27–28, 74–75, 121, 128–29; Brooklyn and, 124–25; Canarsie and, 118, 122; church aid groups and, 116–19; criticisms of, 76–78, 118–22; Eastvillers and, 75–77, 106–7, 128–29; location of, 28–29; role of, 79–80; Westvillers and, 99–112, 115–22
Rieder, Jonathan, 19
Robinson, Cedric, xv
Rockaways, 22–29, 99; Always With You in, 26–28, 80; Black Americans in, 24–26; demographics of, *25*, 26t; history of, 24–25; maps of, *4*, *23*, *25*; Resiliency Is Us in, 79–80; segregation on, 22–27, *25*; social capital loss in, 66–71; social vulnerability index of, *23*. *See also* Eastville; Westville
Russian immigrants, *34*

Sampson, John L., 123
San Francisco earthquake (1906), 77
Saving Grace (org.), 118
Scott, Janny, 25
segregation, 16–19, 22–27, *25*
self-employed people, 37, 134; invisibility of, 78, 89–90; SBA loans for, 56
senior citizens, 74, 105; in boarding houses, 25, 29; informal networks of, 15–16; services for, 71, 73, 138
single-room occupancy (SRO) boarders, 109
slavery, capitalism and, xv–xvi
Small Business Administration (SBA), 55–61, 122
social capital, 5–6, 51, 66–71, 99–101; of basement renters, 39–40, 44; bonding ties of, 102–5;

169

social capital (*continued*)
  infrastructure of, 124; as networking process, 20–21; organizational ties in, 117; racializing processes of, *17*, 17–19; resiliency and, 80. *See also* crisis capital; networking
social ties: deflation of, 21, 67; depletion of, 73, 135; fragmentation of, 84, 134
social vulnerability, 14, 53, 87; chronic, 78; index of, *23, 32*
Solnit, Rebecca, 77–78
spatial inequality, 16
Staten Island, 118
Supplemental Nutrition Assistance Program (SNAP), 53

terrorism, xvi
Tilly, Charles, 22
Tomaskovic-Devey, Donald, 19
transitional housing, 67–68, 91–98

United Methodist Church (Brooklyn), 125–28
urban spaces: crisis capital in, 72–74, 77–78; racializing of, 18–19, 23
utopia, "altruistic," 77–78

Voluntary Organizations Active in Disaster (VOAD), 125

volunteerism, 102–5, 107–8, 118–20; in Eastville, 73–74; organization agglomeration and, 7, 113–21, 128; race/class influences in, 27; in Westville, 115–21. *See also* community-based organizations
vulnerability. *See* social vulnerability

Westville, 100; information diffusion in, 85, 102, 107–8; organization agglomeration in, 116–22; Resiliency Is Us in, 100–113, 116–23; social capital loss in, 70–72. *See also* Rockaways
White flight, 35
White privilege, xviii, xx; ecology of inequity and, 4–5; Pulido on, 16–17. *See also* race/class influences
Wilmers, Nathan, xvii
Wilson, William Julius, 14–15
Winant, Howard, 15
Wooten, Melissa, 19
World Circle (org), 125

zip code tabulation areas (ZTs), 25t, 35t
Zone A, 48, 52
Zone B, 37, 48, *49*

ABOUT THE AUTHOR

Sancha Doxilly Medwinter, PhD, is a sociologist who studies social inequality. She conducts ethnographies of urban and organizational environments. Her research focuses on race, class, and citizen inequality, trauma, and survival among marginalized communities. Dr. Medwinter has held the position of assistant professor of sociology at the University of Massachusetts Amherst for the past five years. She was born in the city of Castries and was raised in Grande-Riviere, on the Caribbean island of Saint Lucia. There, she began her education at Grand-Riviere Combined School. She then secured her placement at the island's top high school for girls, St. Joseph's Convent. She later attended Brooklyn College, at the City University of New York, where she graduated summa cum laude, earning a bachelor of arts degree in political science, with a minor in law and society in June 2009. In 2012, she earned a master of arts degree in sociology from Duke University. She graduated from Duke in May 2015 with a PhD in sociology.

Dr. Medwinter is a public sociologist whose work aims to inform and critique policy, practice, and norms that shape the well-being of economically deprived, racially minoritized, and marginalized communities in the United States and the Caribbean. Her research and teaching areas span race and racism; social inequality; international disasters and crises; poverty, mobility, and social welfare; immigration and citizenship; and Caribbean studies. Dr. Medwinter's most recent publications are "Caribbean Womanism: Decolonial Theorizing of Caribbean Women's Oppression, Survival, and Resistance" (with Tannuja D. Rozario, 2020) and "Reproducing Poverty and Inequality in Disaster: Race, Class, Social Capital, NGOs, and Urban Space in New York City after Superstorm Sandy" (2021).

www.ingramcontent.com/pod-product-compliance
Lightning Source LLC
Chambersburg PA
CBHW031834230426
43669CB00009B/1343